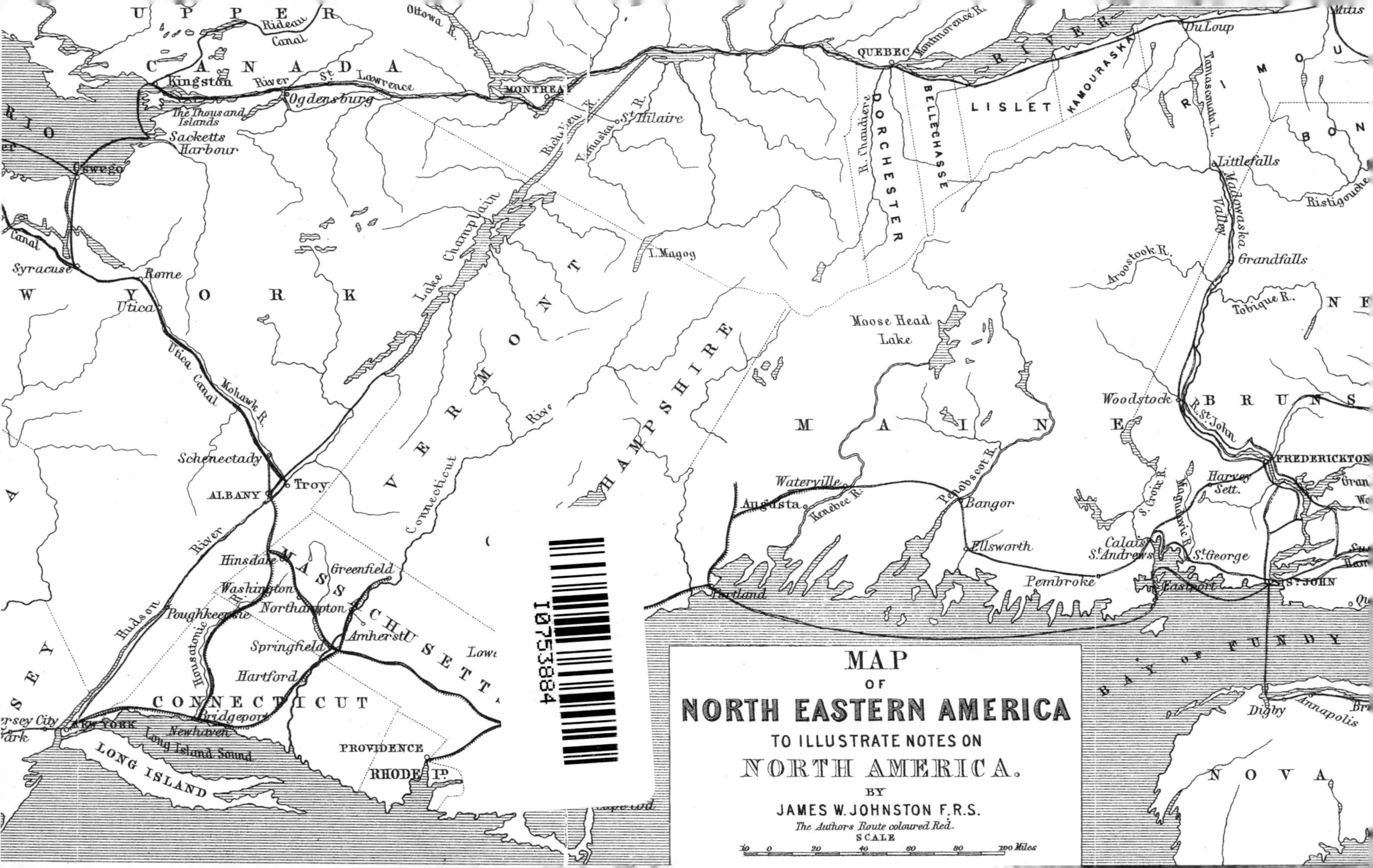
MAP
OF
NORTH EASTERN AMERICA
TO ILLUSTRATE NOTES ON
NORTH AMERICA.
BY
JAMES W. JOHNSTON F.R.S.
The Authors Route coloured Red.
SCALE
10 0 20 40 60 80 100 Miles
UPPER CANADA
NEW YORK
VERMONT
NEW HAMPSHIRE
MAINE
MASSACHUSETTS
CONNECTICUT
RHODE I^D.
LONG ISLAND
NEW BRUNSWICK
NOVA
BAY OF FUNDY
RIMOUSKI
KAMOURASKA
LISLET
BELLECHASSE
DORCHESTER
QUEBEC
MONTREAL
Kingston
Ogdensburg
Oswego
Syracuse
Rome
Utica
Schenectady
ALBANY
Troy
Hinsdale
Greenfield
Washington
Northampton
Amherst
Springfield
Hartford
Bridgeport
Newhaven
Poughkeepsie
NEW YORK
Jersey City
PROVIDENCE
Augusta
Waterville
Bangor
Ellsworth
Pembroke
Portland
Calais
St Andrews
St George
Eastport
ST JOHN
FREDERICKTON
Harvey Sett.
Woodstock
Grandfalls
Littlefalls
Madawaska Valley
Du Loup
Digby
Annapolis
St Hilaire
Moose Head Lake
L. Magog
Lake Champlain
Long Island Sound
The Thousand Islands
Sacketts Harbour
River St Lawrence
Rideau Canal
Ottowa R.
Utica Canal
Mohawk R.
Hudson River
Connecticut River
Housatonic R.
Kennebec R.
Penobscot R.
Aroostook R.
Tobique R.
R. St John
S. Croix R.
Magaguadavic R.
R. Chaudiere
Montmorenci R.
Richelieu R.
Yamaska R.
Tamasconata L.
Ristigouche

NOTES

ON

NORTH AMERICA

AGRICULTURAL, ECONOMICAL, AND SOCIAL

JAMES F. W. JOHNSTON

M.A., F.R.S.S.L. & E., F.G.S., C.S., &c.

READER IN CHEMISTRY AND MINERALOGY IN THE UNIVERSITY OF DURHAM

TWO VOLUMES

VOL. I.

BOSTON:
CHARLES C. LITTLE AND JAMES BROWN

AND

WILLIAM BLACKWOOD AND SONS
EDINBURGH AND LONDON
MDCCCLI

TO

HIS EXCELLENCY

SIR EDMUND HEAD, BART.,

LIEUTENANT-GOVERNOR OF THE PROVINCE OF NEW BRUNSWICK,

&c. &c.

DEAR SIR EDMUND,

I dedicate these volumes to you, partly because they contain, among other matters, the observations I made during a lengthened Tour through the province of which you are the Governor, and in the prosperity of which you feel so lively an interest. But I do so chiefly as a mode of testifying the respect and regard I entertain for yourself and your family, and as affording me an opportunity of expressing my sense of the many acts of kindness I experienced during a prolonged stay at Fredericton, under your hospitable roof.

Believe me, DEAR SIR EDMUND, with great respect,

Your obedient Servant,

JAMES F. W. JOHNSTON.

PREFACE

I HAVE given to the following volumes the title of "NOTES," because I am conscious of the imperfect and hurried character of some of the observations they contain, and that mistakes, generally trivial I hope, and always unintentional, may be discovered in them by natives of North America.

In recording my remarks and impressions, while I am sensible that I have regarded objects with the eyes and feelings of a "Britisher," and have generally written as if I were addressing British readers only; yet I have endeavoured to speak fairly and with candour, both of the institutions and of the social condition of the States and Provinces through which it was my fortune to travel. While I have expressed my opinions freely, I have endeavoured to avoid either ridicule or causeless reproach. And although I cannot hope that my remarks will be always agreeable to my friends in the United States, yet I hope none will accuse me of a desire either to violate confidence, or to return bitterness of speech for the respect and kindness which I everywhere experienced.

In addition to the matters usually commented upon by those who visit foreign countries, the reader will find in these volumes a kind and class of observations which he will not have met with in other books of travels. And though I may appear to incur the risk of injuring their popularity with the general public by introducing agricultural remarks, yet, in the present condition of our own agricultural interest, there are few persons to whom some information in regard to that of America will not be acceptable. These observations on rural matters are also so mixed up with remarks on other subjects as not, I hope, to fatigue even the ordinary readers of books of travels.

It has long struck me as a vital defect in the accomplishments of most of our travellers in foreign countries, that the want of an agricultural eye has prevented them from giving us any of that positive and matter-of-fact information upon which alone a correct estimate of the real character, capabilities, and future economical prospects of a country can be safely based.

I have been more detailed in my remarks upon the lower St Lawrence and the province of New Brunswick, because this is almost untrodden ground, and, so far as I am aware, we possess, in reality, no good account of this region by an eye-witness from

Great Britain. In the province of New Brunswick I spent four months, and travelled two thousand miles—penetrating to the confines of the settled land in nearly every direction. I owe it to the province, therefore, to make its own inhabitants, not less than those of Great Britain and of the United States, better acquainted with the real character and capabilities of its surface. In this respect, I believe the following pages will form a historical document to which future provincial antiquaries will turn back for a description of the state of their country in the middle of the nineteenth century.

Some persons in the United States, and perhaps not a few at home, may be inclined to controvert the opinions I have expressed in regard to the agriculture and to the productive capabilities of the wheat regions of North America. I will not maintain that more knowledge might not somewhat change my views on these subjects; but as these form in reality one of the points in my book upon which I have bestowed much deliberation, I have not put them upon paper without being fully satisfied that they are substantially correct. It will not alter these opinions, that some American writers may dissent from them. My own experience has shown me, that the areas in regard to which individuals in the United States possess really correct and precise agricultural information are very local and limited; while the majority are insensibly inclined to give faith to exaggerations upon this as upon other topics, provided their tendency be the patriotic one of exalting the greatness of their country.

I trust, however, that even where my observations do not wholly coincide with those of my American readers, they will at least acquit me of picking out deficiencies even in their agriculture, for the mere sake of finding fault, or of exposing them in a censorious spirit. I have spoken of the soil, and its treatment, as I would if I were describing a district of Great Britain ; and where I have pointed out defects in past or present practice, it has been for the purpose of mentioning along with them the remedies for past mismanagement, and the improvements of which existing methods are susceptible.

If I may rely upon the testimony of my numerous Transatlantic friends, my temporary residence in New-Brunswick, New England, and the State of New York, has not been without beneficial results to the agriculture of those countries. I trust that, while these volumes make my own countrymen better acquainted with these interesting regions, they will be found to contain not a few hints which may still further benefit and encourage the rural industry, both of Great Britain and of North America. I hope, also, that the general spirit which pervades them will tend to draw still closer the numerous bonds by which our kindred nations are already so intimately allied.

DURHAM, *February* 1851.

CONTENTS

CHAPTER I.

A WEEK IN NOVA SCOTIA.

CHAPTER II.

UP THE RIVER ST JOHN IN NEW BRUNSWICK—FROM THE CITY OF ST JOHN TO THE GRAND FALLS.

CHAPTER III.

UP THE ST JOHN TO LITTLE FALLS, AND ACROSS THE PROVINCE FROM FREDERICTON TO MIRAMICHI.

CHAPTER IV.

FROM THE MIRAMICHI RIVER TO THE CITY OF ST JOHN, BY SUSSEX VALE.

CHAPTER V.

FROM ST JOHN IN NEW BRUNSWICK TO SYRACUSE IN WESTERN NEW YORK.

CHAPTER VI.

THE CITY OF SYRACUSE AND ITS NEIGHBOURHOOD.

CHAPTER VII.

FROM SYRACUSE TO BUFFALO AT THE FOOT OF LAKE ERIE.

CHAPTER VIII.

BUFFALO AND THE NORTH-WESTERN STATES.

CHAPTER IX.

FROM BUFFALO TO THE FALLS, AND DOWN LAKE ONTARIO TO KINGSTON IN LOWER CANADA.

CHAPTER X.

KINGSTON AND ITS NEIGHBOURHOOD.

CHAPTER XI.

MONTREAL AND ITS NEIGHBOURHOOD.

CHAPTER XII.

FROM QUEBEC DOWN THE ST LAWRENCE TO THE MITIS RIVER.

CHAPTER XIII.

THE AGRICULTURE AND WHEAT-PRODUCING CAPABILITY OF THE UNITED STATES AND CANADA—AND THE NAVIGATION OF THE ST LAWRENCE.

CHAPTER XIV.

FROM MITIS ON THE ST LAWRENCE BY THE KEMPT ROAD ACROSS THE PENINSULA OF GASPÉ TO CAMPBELTON AND DALHOUSIE ON THE RESTIGOUCHE.

NOTES

ON

NORTH AMERICA

CHAPTER I.

Halifax in Nova Scotia.—Fresh complexions of the people.—Roman Catholic fête.—Roman Catholics in Halifax.—Precedence and title conceded to Bishops.—Coloured people in Nova Scotia.—Micmac Indians.—Maritime commerce of Nova Scotia, its certain extension. —Mackerel fishery.—Shoals of mackerel.—Export of salt fish.—Scratched rocks, and agricultural character of the neighbourhood of Halifax.—Stony and unfertile surface of the coast line.—Young's *Letters of Agricola*.—Increase of population in Nova Scotia.—Proportion of the agricultural produce to the population.—Inner Bay of Halifax.—Railway from Halifax to Windsor.—Soils and forests of the Ardoise hills.—Drought of 1849.—Pacing horses of Canada.—How trained in Sardinia.—Gypsum quarries at Windsor.—River Avon.—Dyked alluvial lands of the Bay of Minas.—Varieties of land, and their money-values.—Sand plain of Aylesford.—Structure of the vale of Annapolis.—Town of Annapolis.—Ice-holes in the North Mountains.—Ironworks of Bear river.—Healthiness of the country.—Handiness of the Nova Scotians.—Blue-nose provincialisms.

On Saturday the 28th of July, at 3 P.M., I sailed from Liverpool in the steam-ship America. We took the northern course; lost sight of the west coast of Ireland on the afternoon of Sunday the 29th; about noon of the following Sunday came in sight of Newfoundland; and early on the morning of Tuesday the 7th of August, I landed at Halifax in Nova Scotia. We had thus a pleasant passage of nine days and fifteen hours; and as we had an agreeable party, we felt almost sorry our voyage had been so short.

The noble harbour of Halifax, in which all the navies of the world might securely float, is only one of the countless inlets and basins which the Atlantic coast of Nova Scotia, from Cape Canseau to the Bay of Fundy, everywhere presents. The jagged outline of this coast, as seen upon the map, reminds us of the equally indented Atlantic shores of Scandinavia; and the character of the coast, as he sails along it—the rocky surface, the scanty herbage, and the endless pine forests—recall to the traveller the appearance and natural productions of the same European country.

The coast of Nova Scotia is indeed very unpromising in an agricultural sense; and though of the surface of the province there are in reality three and a half millions of acres which present to the Norwegian, the Swede, or the Finlander, the rocky soils, scenery, and, generally speaking, the natural productions of his own country, yet both Nova Scotia and New Brunswick have in reality been unjustly depressed in European estimation by the character of their shores. The greater number of those who have hitherto returned to Europe from this part of North America, and who have regulated European opinion in regard to it, have seen only the coast line, or the interior of its rocky harbours; and these are certainly as naked and inhospitable as an inhabited country can well be. Those who have sailed along the Baltic shores of Sweden and Finland, or to Gothenburg by the estuary of the Gotha, or among the rocks and inlets of the western coast of Norway, will be able to realise, without visiting them, what the sailor sees on the shores of our American colonies.

But the interior parts of these provinces are not represented by these barren borders. Though they do contain large tracts of poor and difficult land, yet rich districts recur at intervals, which rival in natural fertility the most productive counties of Great Britain. The colonists complain, with reason, that the evil opinion entertained of them has diverted the tide of English settlers, English capital, and English enterprise, to more southern or western regions, not more favoured by nature than they are themselves.

A European stranger who, on landing in Halifax, looks for the sallow visage and care-worn expression which distinguish so many of the inhabitants of the northern States of the Union, will be pleased to see the fresh and blooming complexions of the females of all classes, and I may say of almost all ages. Youth flourishes longer here, and we scarcely observe, in stepping from England to Nova Scotia, that we have as yet reached a climate which bears heavier upon young looks and female beauty than our own.

The day of my landing at Halifax was a fête-day among the Roman Catholic schools. Twelve hundred children, in holiday dresses, were marching in long procession by nine in the morning, with flags and banners and music, along the main street of the city, and thence under a triumphal arch of flowers and an avenue of green pine-trees, planted for the occasion, to a steamboat which was in waiting to convey them across the bay to M'Nab's Island, where the amusements of the day were provided. As music and dancing and refreshments were among the entertainments, this fête attracted a large assemblage of all parties, whom rigid religious views did not restrain from countenancing a public display of the Roman Catholic body. As a stranger, I was grateful to the provincial secretary, Mr Howe, for an invitation to accompany him and his family in the afternoon to the scene of festivity. We crossed the bay in a steamboat crowded almost to suffocation; and it was here, and among the thousands whom I saw on the island, that I was enabled to judge of the adaptation of the northern climate to the complexions of our island population.

In Europe, it is in countries which, like Great Britain, Ireland, and Holland, are surrounded by an atmosphere rarely arid or dry, either from excessive cold or from excessive heat, but which, more or less loaded with moisture, always softens and expands the minute vessels of the skin, that health and freshness of complexion in both sexes is most conspicuously perceived and most permanent. To the fogs and rains, therefore, which so frequently visit this and other parts of the North American coast, lying within the influence of the Gulf Stream, the healthy looks of the people are probably in some measure to be ascribed.

I was early struck, on this my first day's residence in North America, with the less constrained and more equal intercourse which appeared to prevail between what we should call the different classes of society. The servant and the mistress, the mechanic and the barrister, with little distinction of dress or behaviour, discoursed on a perfect equality, and persons filling the highest political offices were jostled about as unceremoniously, and were as familiarly hailed, as the humblest of the crowd. The secret is, that every one feels what I understood when my friend said to me, "That girl may marry, and be better off than her mistress to-morrow; and the lowest of these men may rise to the highest civil office in the province." As the ermine of the bench, and the mitre of the archiepiscopal seat, secure to the humblest member of two of our learned professions in England a portion of that respect with which we look upon a future Lord Chancellor or a possible Archbishop, so I suppose the sense of equal opportunities being open to all entitles each man in these provinces to a more equal consideration.

The Roman Catholic body in Halifax is strong and growing, chiefly through the yearly accessions of emigrant Irish and their descendants, who here appear to thrive, and are said to be well-behaved. The Presbyterians used to be, and probably still are, the most numerous of the religious sects in Halifax, and next to them the Episcopalians of the Church of England. The Roman Catholics have of late years increased, and they have obtained an advantage over the non-Episcopal sects in the title of "My Lord," lately conceded to their bishop by order of the Home Government, and in virtue of which he takes rank with the Church of England bishop, and precedence of all the dissenting clergy.

That this is a great grievance in the eyes of the Presbyterians and others, in the two colonies of Nova Scotia and New Brunswick, I was scarcely a day in Halifax till I had learned. Until recently, the bishop of the "Church of England in the colonies" was the only person addressed as "My Lord,"—a solitary and invidious title among a people composed, for the most part, of what we call dissenters in England, and in a country where so little distinction of ranks prevails. It became less singular when the same title was conceded to the Roman Catholic bishops, and, of course, a greater number of persons became interested in keeping up this distinction. But the hostile feeling was in consequence only made stronger in the breasts of the majority of the people.

Such distinctions in a colony, it appears to me, ought to be conceded, not for an imperial, but for a provincial reason—not because a certain religious body is powerful in Europe, but in consideration of the feelings and wishes of a large body of the colonists themselves. Now, if this latter reason had been influential, there are other sects to whom some equal distinction ought to have been conceded. The Presbyterians and Baptists are both stronger bodies than either the English Episcopal or the Roman

Catholic in the colonies of Nova Scotia and New Brunswick, and therefore more entitled to consideration at the fountain of honour. If there be any way in which it could be done, therefore, the head of these several bodies —their moderator or president for the time being— should be equally honoured with the more permanent heads of the Episcopalian sects;—that is, if the distinctive title is to be retained at all, and the precedence of high clerical office retained.

It may be said that the Presbyterian, Baptist, and other bodies, are opposed upon principle to the connection of honorary precedence with clerical office, and have therefore never asked such distinctions for the head of their several denominations. This is probably true; but my intercourse with the inhabitants of these colonies has satisfied me that much lurking ill-will against the mother country has arisen from the kind of half-establishment originally granted to the English Church; and that this ill-will, instead of being lessened, has been deepened in intensity by the selection of the Roman Catholic body for a similar distinction. Why should the mother country procure ill-will—manufacture it, I may say, for herself— by intermeddling in the religious disputes of the different denominations in the colonies? Either we ought to leave these entirely to the control of the local legislature, as all other internal political and social matters now are left, or the offer, at least, of similar honours should be made to the head of each religious body possessing a certain numerical force, and consequent political weight, in the province. This offer, whether accepted or not, would at least remove the complaint of invidious distinctions from the shoulders of the Home Government, and would confine the discussion in future to the general question of precedence or no precedence to the holders of high clerical office. Should any unfortunate circumstances bring about a separation from the mother country, such distinctions in Nova Scotia and New Brunswick would certainly not be permitted to exist for another hour.

On the field upon M'Nab's Island, where the people were assembled, were music and dancing parties in different places; swings and refreshment stalls, whites of all grades, and *darkies* of different shades; but I saw neither intoxication nor disorder, nor rudeness, nor incivility anywhere. A little of the liveliness of the early French settlers probably clings to the modern Nova Scotian; but though there were many both Irish-born and of Irish descent among the crowd, there was no shade of a disposition to an Irish row.

Many coloured people, some apparently full-blood negroes, were to be seen in the streets of Halifax acting as porters, and in other humble employments. A few of these looked miserable enough.

As far back as the close of the American war, numbers of coloured people came here, either with their loyalist masters, or alone, and at the expense of Government. These early settlers have multiplied and become to a certain extent acclimatised, and many of them are industrious owners of small farms. Generally, however, the negroes are spoken of as indolent, as hanging about the towns, and as suffering much from the severity of the winter.

People of colour enjoy, I believe, in all the British colonies of North America, the same political privileges as are possessed by other classes of her Majesty's subjects. I went into the jury court, where the author of *Sam Slick* was the presiding judge, and I was both surprised and pleased to see a perfectly black man sitting there in the box as a juror.

Among the other novelties to a stranger in Halifax is an encampment of the Micmac Indians, whose wigwams I found pitched upon some high ground above the town of Dartmouth, on the opposite side of the bay. These Indians have a broad Asiatic face; and are more intelligent, but less patient of restraint than the negroes. Little real success has attended the many attempts which have been made to educate and localise them. They have become faithful Roman Catholics, are obedient to their priests, regular at confession, and very honest; but they do not settle steadily to the monotonous labours of agriculture, or to the confinement either of domestic service, or of regular handicraft or mechanical trades.

In the first wigwam I entered, I found half-a-dozen men playing at cards; and, in the next, as many women and children making baskets. Their English is broken, and to each other they converse in their native tongue. They are diminishing in numbers, many having been carried off some time ago by a fever, which raged specially among themselves; but there are said still to remain five thousand of them in Nova Scotia.

In the harbour of Halifax, I saw few large ships; there were, however, many small vessels employed either in the fisheries or in the coasting trade to the States and the Canadas. There are four circumstances which seem to concur in promising a great future extension to this maritime portion of Nova Scotian industry. In the first place, the sea and bays, and inlets along the whole Atlantic border, swarm with fish of many kinds, which are the natural inheritance of the Nova Scotian fishermen. Second, this coast is everywhere indented with creeks and harbours, from which the native boats can at all times issue, and to which they can flee for shelter. Thirdly, there exists in the native forests—and over three millions of acres in this province probably always will exist—an inexhaustible supply of excellent timber for the shipbuilder. And, lastly, from the influence of the Gulf stream most probably, the harbours of Nova Scotia are, in ordinary seasons, open and unfrozen during the entire winter; while, north of Cape Canseau, the harbours and rivers of Prince Edward's Island and of the Canadas are closed up by ice. This latter circumstance, if a railway should be made from Halifax to the St Lawrence, ought to place the West India trade of a large portion of the Canadas and of New Brunswick in the hands of the Nova Scotia merchants — while all the circumstances taken together will doubtless, in the end, make them the chief purveyors of fish both to Europe and America. At present, they complain of the bounties given by their several Governments to the French and United States fishermen. But bounties are in all countries only a temporary expedient: one part of a people gets tired at last, of paying another part to do what is not otherwise profitable; bounties are therefore abolished, and employment in consequence languishes. The fisheries of Nova Scotia are the surer to last that they are permitted or encouraged to spring up naturally, without artificial stimulus, and in the face of an ardent competition.

Of the coast fisheries, the most important to the trade of Halifax is that of mackerel. This fish abounds along the whole shores, but the best *takes* are usually made in the Gulf of St Lawrence, off the shores of Cape Breton and Prince Edward's Island, and especially at Canseau, where the quantity of fish has been "*so great at times as actually to obstruct navigation.*"* The excitement caused by the arrival of a shoal of mackerel, is thus described by Judge Haliburton, in *The Old Judge*:—

"Well, when our friends the mackarel strike in towards the shore, and travel round the province to the northward, the whole coasting population is on the stir too. Perhaps there never was seen, under the blessed light of the sun, anything like the everlasting number of mackarel in one shoal on our sea-coast. Millions is too little a word for it; acres of them is too small a tarm to give a right notion; miles of them, perhaps, is more like the thing;

* Gesner's *Industrial Resources*, p. 124.

and, when they rise to the surface, it's a solid body of fish you sail through. It's a beautiful sight to see them come tumbling into a harbour, head over tail, and tail over head, jumping and thumping, sputtering and fluttering, lashing and thrashing, with a gurgling kind of sound, as much as to say, 'Here we are, my hearties! How are you off for salt? Is your barrels all ready?—because we are. So bear a hand and out with your nets, as we are off to the next harbour to-morrow, and don't wait for such lazy fellows as you be.'" *

A ready market for this fish is found in the United States; and the absolute as well as comparative value of the trade to Nova Scotia, may be judged of from the following return of the quantities of pickled fish of the most plentiful kinds, exported from Halifax in 1847:—

	Barrels.
Alewives, . . .	7000
Salmon, . . .	6000
Herrings, . . .	22,000
Mackerel, . . .	190,000

From Cape Breton and Newfoundland the largest export consists of cod-fish.

The day after my arrival at Halifax, I drove round the peninsula on which the city stands, and up the northwest arm—an inlet or creek, by which the peninsula is formed, and which runs inland from the bay a few miles behind Halifax.

To one who wishes to form a general idea of the agricultural character and capabilities, as well as of the geological structure and botanical relations of the Atlantic border of the province, this drive is very instructive. On a clear sunny day the views are beautiful, and the ride most exhilarating. The old slate rocks are interspersed with masses of granite—probably, in many cases,

* *The Old Judge*, by Sam Slick, vol. ii. p. 96.

only old stratified rocks, a little more changed than themselves—while stunted pine-woods and peaty hollows form the principal features of the surface. Anciently submerged, however, as all this country has been, there are everywhere visible traces of those currents or glaciers which about the same period scratched and grooved so large a portion of the northern continents of Europe and America. Scratches, continuous, deeply cut, generally parallel, but frequently crossing each other at angles of ten to twenty degrees, are beautifully seen on the broad naked granite surface of Point Pleasant, on which the fort stands, upwards of a hundred feet above the sea, and at other places in that immediate neighbourhood. These markings, with the accumulated drift and boulders, strengthen more the general likeness of the country to what the visitor may have seen about Stockholm in Sweden, or Helsingfors in Finland.

Difficult to the farmer, and eminently stony, the country about Halifax really is. In some places, boulders of various sizes are scattered sparsely over the surface; in others they literally cover the land; while in rarer spots they are heaped upon each other, as if intentionally accumulated for some after use. One ought to visit a country like this, while new to the plough, in order to understand what must have been the original condition of much of the land in our own country, which the successive labours of many generations have now smoothed and levelled.

When Cæsar invaded Britain, stony deserts might exist where the plough now easily cuts the soil; so that the greater produce is not due alone to the higher skill of those who now cultivate the land, but more probably to the effect of labour and hard toil expended upon it by drudging serfs in former ages. The northern end of Lough Corrib, in Ireland, would probably still bear a comparison with many of these difficult places in North America. The huge walls of stones which the peasantry have gathered from their fields in other parts of the same island, indicate that, within comparatively recent periods, they have been little better; while what England has been may be inferred from the fact that, in an old-farmed district in Northumberland, I have myself known of six hundred cart-loads of trap boulders being raised and carried out of a single field. I am less inclined, therefore, than some may be to bewail as hopeless the apparently unimproveable condition even of the stonier parts of Nova Scotia and New Brunswick. The progress of agriculture in such districts is necessarily slow, but a thousand years will do for these countries infinitely more than it has done for us. Productive fields and farms have indeed already risen in many places from among the rocks and stones around the city of Halifax. The market it affords for produce, and the wealth from time to time accumulated by its merchants, have had their effect upon the surface; and gardens and fields and small farms have gradually spread their cheerful surfaces along the hilly slopes which skirt the beautiful bay. But where and while such stony tracts occur, arable farming on a large scale can never be carried on. It is not in this neighbourhood, therefore, that the agricultural emigrant is to look for those rural attractions which are to dispose him to settle in Nova Scotia.

One would scarcely expect that much should ever have been done in such a locality for the general promotion of North American agriculture. And yet I was much interested to meet with a work published at Halifax in 1822, under the title of *Letters of Agricola*, by John Young, Esq.—the father, I believe, of the present Attorney-General of the province—which, for sound knowledge of the subject, both practical and scientific, for honest common sense, and for a warm but prudent zeal to improve the country in which he lived, is, as a whole, superior to any other book of the time I have hitherto met with in any language. It was not to be wondered at that, through the exertions of Mr Young, a provincial Board of Agriculture should have been established, and many county agricultural societies, which still exist, though less patriotically urged forward, perhaps, than in his time.

The publication of the *Letters of Agricola* marks an era in the agricultural history of the province; the writings of the author of *Sam Slick* an era, not only in its social history, but in that of the steam traffic and intercourse of the world. Both writers must rank among the truest patriots of Nova Scotia. Is there none in the province now who can take up the mantle of Young again, and re-awaken, in behalf of agriculture, the spirit which, thirty years ago, when less was known of its principles, he was so successful in creating?

If we are permitted to draw any conclusion from the increase of population in Nova Scotia, this province would appear to have advanced as rapidly as almost any other part of North America. The number of its inhabitants, at different periods, is stated to have been—

In 1772,	18,300	In 1826,	130,000
... 1781,	12,000	... 1846,	280,000
... 1784,	32,000	... 1850,	300,000

The province has many resources in fishing, mining, and agriculture, and cannot be prevented from increasing, both in population and in wealth. But its progress will be more rapid in proportion to the wisdom, energy, and singleness of purpose of those whom the colonists—to whom all public officers are now responsible—may select to manage their affairs.

It possesses an area of nine and a half millions of acres, of which five and a quarter millions are granted to private parties, and four and a quarter still remain in the hands of the provincial Government. It does not grow corn enough for its own consumption; but Dr Gesner states, that not a fiftieth part of the surface is cleared of timber, and that not a hundredth part is in cultivation.*

Now, one hundredth part of the whole area is about 95,000 acres; and supposing this to produce, at the same rate as the cultivated land of Great Britain—of which each 170 acres supports 100 inhabitants—they would raise food for about 60,000 inhabitants. But the surface of Nova Scotia is not so well cultivated or so productive, as a whole, as Great Britain. Its 95,000 cultivated acres, therefore, do not support so many as 60,000 of its people. On the other hand, it is certain, from the quantity of food actually imported, that more than 60,000, or one-fifth of the population, must be maintained by what the province itself produces. I conclude, therefore, that Dr Gesner's estimate of the proportion of the province which has already been brought into cultivation is largely understated.

Again, according to the estimate of Dr Gesner, not more than one-half of the population is employed in agriculture, the rest being engaged in lumbering, fishing, &c.† That is, each person employed in agriculture raises less food than is necessary to support two people—since there is a large importation of American flour. But, in England and Scotland, only one-fourth of the population is engaged in, or dependent upon, agricultural employment; that is, each person occupied in tilling the land raises food for four people. Hence Nova Scotia is not made to yield half so much food as Great Britain, in proportion to the number of people employed in agriculture, if Dr Gesner is nearly right as to the number so employed in Nova Scotia.

I think he can scarcely be under the truth in estimating the agricultural population at one half of the whole. We are compelled, therefore, to conclude, either that the land in general is not grateful for the labour expended upon

* *Industrial Economy of Nova Scotia*, p. 23. † *Ibid.*, p. 24.

it, or that the inhabitants are deficient in industry; or that, from want of agricultural skill, their labour is not turned to the best account, and the capabilities of their soil not fully brought out.

From my brief stay in the province, and the peculiar aridity of the season, I had not an opportunity of determining these points by my own observation; but, from all I have learned, I am inclined to attribute much of the comparative deficiency of produce in the province to a want either of skill or of persevering industry on the part of the cultivators.

On the morning of Thursday the 9th of August I left Halifax, by stage, for Windsor, whence the steamboat proceeds across the Bay of Minas and the Bay of Fundy, to St John in New Brunswick. The morning was fine, and the ride up the west side of the bay very delightful. The harbour of Halifax consists of an outer and an inner bay, both of great extent. The inner bay, which is completely land-locked, is as yet little frequented, even by boats; and one laments to see so many fine sites for houses and clearings, on both shores, unoccupied and almost desolate. The land in general is poor and stony, and wealth has not yet so largely accumulated at Halifax as to give a value to the comparatively unproductive margins of this wide inner lake.

A ride of ten miles brought us to Sackville, at the head of the lake, where we stopped to breakfast. In passing over these first ten miles on a new continent, a native of Great Britain or Ireland, though not learned in trees, can hardly fail to be struck with the new and varied general outlines, forms of leaf, and appearances of bark, which force themselves upon his attention. The varieties of pines, maples, and birches, and the peculiar foliage of the native oak and ash trees, give much novelty to the journey along the borders of this lake. Little spots of land occurred along its margin, of various quality, and suited, therefore, to the growth of a larger number of species of trees than are usually seen over much greater distances in the interior of the country.

Above Sackville, which as yet is little more than an inn, with its necessary outbuildings—but which will, no doubt, be the site of a future town—several small streams unite in a main valley, and empty themselves into the head of the lake. Up this valley, and across the peninsula, runs the road to Windsor and the line of the projected railway between Halifax and that town. As a means of facilitating and hastening the communication between Nova Scotia and New Brunswick, the execution of this railway would be of great benefit to both provinces. With good steamers on the Bay of Fundy, this short line would reduce the distance in time between Halifax and St John to ten or twelve hours, and would make the Cunard steamers greatly more useful to the province of New Brunswick. And if, as is likely to be the case, the transport of merchandise across the Atlantic by steam is to be diminished in cost, and to become much more common and extended, this railway would at once promote the extension of such a traffic. And while by this traffic the steamers themselves would be aided, the commerce of New Brunswick would obtain the advantage of a direct and easy weekly access to the European markets. Were this railway constructed, we might hope, within a brief period, to see merchant steamers plying between Halifax only and the British ports on the Mersey, the Clyde, and the Bay of Galway, laden with the lighter articles of traffic which the merchants on either side can mutually interchange. Thus, not only would the settlement and more rapid improvement of the part of Nova Scotia, through which the railway is to run, be greatly promoted, but the general commerce of the two provinces also increased, and the extension and profit of Atlantic steam communication hastened on.

Poor clay-slate soils, thin and cold, with occasional quartz rock, and granite in mass or in drifted boulders, accompanied us beyond the summit of the Ardoise hills, which form the water-shed between the Atlantic rivers and those which empty themselves into the Bay of Minas. Pine forests mostly usurped the surface, though here and there, on the margins of lakelets, or where flatter and less stony tracts occur, labour and industry had overcome nature, and compelled rich herbage and moderate corn to spring up in their stead. It was interesting to observe how the absence of human labour for a few years gave again uncontrolled supremacy to the natural vegetation; and pine forests, young, but flourishing and dense as ever, gradually covered again even long-established clearings.

The summer and autumn of 1849 will long be remembered in the British provinces of North America, as well as in the north-eastern States of the Union, for its excessive drought. The first striking effects of it I had yet seen came under my observation to-day, in the burnt forest we occasionally passed on either side of the road, and in the blazing trees and underwood, which, in a few places, hemmed us in on both sides, and, with horses less accustomed to fire, might have proved a source of danger. It was remarkable to see how much the soil, and the seeds it contained, seemed to have been quickened by the passage of the fire. Ferns and fire-weeds embraced the blackened stumps and trunks of fallen trees, while smoke still lingered around them; and I was assured that a couple of weeks was often sufficient to produce such effects.

After crossing the water-shed, which rises about seven hundred feet above the sea, and descending about halfway on the other side towards Windsor, we left the stony, granite, and metamorphic slates, and entered upon soils of a more propitious character, derived from those gypsum-bearing and red sandstone rocks which have been referred to the lower part of the Nova Scotia coal formation. The country is also more undulating, better inhabited, more generally cleared—bearing corn or useful herbage—and has a less humid and changeful climate than the Atlantic slope of the Ardoise hills. Here I first saw a field of growing Indian corn; and, as we stopped to change horses, had an opportunity of walking into and examining it. But I could not repress a feeling of melancholy as we drove along, and saw vegetable life everywhere suffering from the excess of drought. Herbage for the cattle was scarcely to be obtained; the grass fields were burned up, and displayed one universal brown. The hay crop had almost entirely failed, and how to obtain winter food for the stock had already become a matter of most difficult consideration. The reader who is possessed of an agricultural eye will judge how far it was possible for a stranger passing through it, to form, under such circumstances, an idea of the agricultural capabilities of the country. I afterwards saw much of the same effect of drought in New Brunswick and the north-eastern States; and I was informed by those who had known the province for forty years, that nothing equal to the drought of 1849 had been experienced in their time.

On starting with our new team of horses, my attention was arrested by the peculiar gait of the off leader. It slipped and waddled along, alternately lifting and resting upon the fore and hind feet of the same side, a pace I had never seen before. It proved to be a Canadian horse, trained, as they frequently are in that province, to this peculiar pace. It is a sort of shuffling, awkward-looking gait, but is very easy for riding. It is said that a person may ride a whole day at this pace without any fatigue. I hoped to have been able during my subsequent visit to Canada to make a trial of this alleged easiness to the rider, but the opportunity did not fall in my way. Horses so trained are known as *pacing* horses, and the practice has probably been introduced by the French settlers.

I have never myself seen it in France, and should suppose it to be an uncommon pace even there, and that it has most likely been introduced from the shores of the Mediterranean. I find a notice of it in a work upon Sardinia, lately published by Mr Warre Tyndale.* "Much attention," he says, "is paid to giving the better class of horse a peculiar step called *portante*, for which we have neither a corresponding word or pace, being something between an amble and a trot, and taught in the following manner:—

"The fore and hind legs are attached to each other by two cords, supported by others fastened to the saddle so as to prevent their dragging on the ground; and, thus fettered, the horse is put in action—the trainer pulling the right and left side of the bit, alternately, and giving a corresponding pressure with his leg, which forces the animal to move either the two off or the two near legs simultaneously, producing thereby an easy *glissade* step. It has been compared to the Turkish amble, but, judging from personal experience, it is as dissimilar as it is to our cavalry or farmer's trot. The movement is delightfully easy, especially easy where one has to be on horseback for many consecutive hours; and, as Cetti says, 'Il viaggiare in Sardegna e perció la piu dolce cosa del monde: l'antipongo all' andare in barca col vento in poppa.' The travelling in Sardinia is, on this account, the most agreeable thing in the world: I prefer it to going in a boat with the wind astern."

I do not know how the training is effected in Canada, but it is very interesting to find this pace prevailing in two countries so remote from each other. May it not have been introduced into Canada by some of the

* *The Island of Sardinia.* London, Bentley, 1849, vol. i. p. 200.

Romish clergy from the islands or borders of the Mediterranean?

Windsor, which we reached after another hour's drive is a neat, clean, well-built little town, standing on the estuary of the Avon, and within a short distance of the mouth of the St Croix river. Both of these rivers empty themselves into the Bay of Minas, and are distinguished by the lofty white cliffs of gypsum which are seen at various places along their banks. The country adjoining the lower part of both rivers is in many places gypsiferous, and the undulating appearance of its surface, the rounded hills, and the sudden hollows which here and there appear, are in great part to be ascribed to the numerous swallow holes and sinkings which have been produced through the gradual solution and removal, by surface water or by springs, of the gypsum from beneath. A similar surface of rounded hills and hollows afterwards attracted my attention along the shores of the Cumberland basin, in some parts of New Brunswick, and on the gypsiferous strata along the out-crop of the upper beds of the Onondaga salt group, and the base of the Helderberg limestone in Western New York.

After a hasty dinner, at the small but clean town of Windsor, I paid a hurried visit to the plaster quarry of Judge Haliburton, which affords the principal article of export from the river Avon. The gypsum occurred and was worked very much as our limestones are, forming a face of rock in which different layers were visible of various degrees of whiteness, and crystalline structure. The whitest and purest is quarried and conveyed, by an economical railway to the river, where it is shipped chiefly for the United States.

At Windsor, it is usual to embark in the steamer for St John in New Brunswick. In favourable weather this is a run of twelve or fourteen hours with the steamers now on the station. That I might see a portion of the richest land in the province, however, I had been recommended to proceed westward to Annapolis, about eighty miles by land, and thence by a steamer which plies regularly to the city of St John.

Starting again with the stage, we ascended the Avon till it became sufficiently narrow to be bridged over, and then crossed to Falmouth by one of those covered wooden bridges of which I afterwards saw so many in North America. They form long dark wooden tunnels, stronger, perhaps, and more durable for their darkness, but most effectual in preventing either the beauties or defects of the river scenery from reaching the eye of the passenger.

Whoever has sailed up the Avon to our English Bristol when the tide was low, would, this afternoon, have agreed in the propriety of the name which has been given to this river of Windsor. The tide was low, and, as in the English Avon, lofty and steep mud banks confined the waters, and showed at once how high the tide must rise, and how fertilising its muddy water must be.

From this point the land had an improved appearance, and the first good crop I had seen during my whole day's ride began to cheer my eyes. As we drove along, I gradually shook off the feeling of despondency, with which I had looked upon the parched upland country through which I had come to Windsor. I was now proceeding over a more elevated and less valuable portion of that rich alluvial land, for which the shores of the Bay of Minas, and its tributary creeks, and of the head-waters of the Bay of Fundy in general, have been long famous. Advancing twelve or fifteen miles further to Horton and Wolfville, I found myself on the edge of the richest dyke-land in the province. I quitted the stage at Wolfville, for the purpose of taking a drive over a portion of the most productive land before the evening set in.

Having obtained a light carriage, and an intelligent guide, I drove over some dozen miles of what is certainly a naturally fertile, and was then a comparatively smiling, district. But even at this low level, and so near the waters of the broad bay, the drought had seared and yellowed the usually luxuriant herbage; and had I not come from a far more arid region, it would have conveyed to my mind the impression that the agricultural capabilities of the township of Cornwallis had been much over-estimated.

These dyked alluvial lands of the Bay of Fundy are to Nova Scotia and New Brunswick what the carses of Gowrie and Falkirk are to Scotland, and the warped lands of Lincolnshire to Eastern England. The thick waters of our Humber and Trent give a fair idea of those of the Bay of Minas, and of the other broad creeks which communicate with the Bay of Fundy; only the American waters are scarcely so dark in colour, and the mud they deposit is of a redder hue. The frequent villages, and the numerous scattered habitations, which are visible from the higher ground of Cornwallis, are abundant proof of the productiveness of the soil of this more favoured part of Nova Scotia. It is not all, however, of equal quality.

Three kinds of land are distinguished in this and the adjoining province. *First*, dyke-land—the rich alluvial deposit of these waters, so called from its having been laid dry by a succession of dykes, which for the last two centuries have been gradually advancing beyond each other towards the bay. This land sells at present at from £15 to £40 sterling per acre; and some of it has been tilled for 150 years without any manure—a treatment, however, of which it is now beginning seriously to complain. It averages 300 bushels (9 tons,) and sometimes produces 600 bushels (18 tons,) of potatoes to the acre. *Second*, Intervale—the generally light alluvial soil, which in variable width fringes the banks of the rivers above the head of the tide-waters. This also varies in quality, but with farm-buildings is rarely valued so high as £20 an acre. *Third*, Upland — elevated above both rivers and tides, and which owes nothing to either. Over a large portion of the province, the upland is said to be comparatively fertile, and free from stones. The most improved of this kind of land, however, with farm-buildings attached, rarely sells so high as £10 an acre. The wild or wilderness land is granted by Government at about 3s. 6d. an acre.

The Baptists, as I have already observed, are a powerful body in these provinces. At Wolfville, they have a college or academy, attended by a large number of students. It is a handsome building, situated on a rising ground, which overlooks the rich flats beneath, the Minas basin beyond, and carries the eye over to the Cobequid Mountains on the other side of the sea. Before reaching Windsor, we passed, at a short distance on our left, a Church of England college, also finely situated, but said not to be so well frequented, or in so flourishing a condition, as its friends would desire.

By starting early in the morning, I was enabled to advance as far as Kentville before the departure of the stage, and to proceed along the valley to Annapolis, a distance of nearly seventy miles. The road, in general good, though in some places sandy, runs along the foot of what are called the South Mountains, from their skirting this long valley on the south. It rises very gently and very slightly till it reaches an immense bog—called in these provinces a Carriboo bog or Carriboo plain—which is the water-shed from which flow both the Cornwallis river and that of Annapolis, in opposite directions; thence it descends as gently to the town of Annapolis.

Along the lower part of each river there is much good land, but towards the middle of the day's journey, especially about Aylesford and after passing the bog, it becomes sandy; and there is here, occupying a large breadth of the valley, an extent of many miles of light and comparatively worthless land. On this poor soil I saw, for the first time, the sweet fern, *Comptonia asplenifolia*, which I became well acquainted with in my after journeys in New Brunswick. It rejoices in light, sandy, almost useless soils, of which I know scarcely any more sure practical indicator.

The "Old Judge" thus describes what he calls the great Aylesford sand-plain :—

"The great Aylesford sand-plain folks call it, in a ginral way, the Devil's Goose Pasture. It is thirteen miles long and seven miles wide; it ain't jest drifting sand, but it's all but that, it's so barren. It's oneaven, or wavy, like the swell of the sea in a calm, and is covered with short, thin, dry, coarse grass, and dotted here and there with a half-starved birch and a stunted mis-shapen spruce. Two or three hollow places hold water all through the summer, and the whole plain is criss-crossed with cart or horse tracks in all directions. It is jest about as silent, and lonesome, and desolate a place as you would wish to see. Each side of this desert are some most royal farms—some of the best, perhaps, in the province—containing the rich lowlands under the mountain; but the plain is given up to the geese, who are so wretched poor that the foxes won't eat them, they hurt their teeth so bad. All that country thereabouts, as I have heard tell when I was a boy, was oncest owned by the lord, the king, and the devil. The glebe-lands belonged to to the first, the ungranted wilderness-lands to the second, and the sand-plain fell to the share of the last, (and people do say the old gentleman was rather done in the division, but that is neither here nor there,) and so it is called to this day the Devil's Goose Pasture." *

* *The Old Judge*, vol. ii. p. 5.

It is a pleasant thing in the British provinces to hear the people talk of England and Scotland and Ireland—of the Old Country generally—as *home;* and it is pleasant to meet so many persons who, though long settled, and having families of province-born children, were themselves born at home, and like to ask of places they knew in their youth from one who has lately seen them, and to tell how they have struggled and fared since they came to the New World. Those persons are greatly deceived who think that less labour, and less patience and perseverance, are necessary to success in the New World than in our part at least of the Old. The chief difference is, that there is room enough in the broad lands of America for the full employment of all, and that the diligent man of moderate desires is sure of a competency.

Along this road I met the first examples of these old settlers, and I was especially interested by the narrative of an old Aberdonian, at whose house we stopped to refresh our horses. He had remained fixed where he first settled, and the determination he brought with him from his native country had at length made him master of almost everything desirable around him.

As we descended towards Annapolis, the land and country improved, and the last fifteen miles were beautiful in scenery, and showed extensive fertile flats in the bottom of the valley. Bridgetown, ten or twelve miles above Annapolis, struck me forcibly as neat, clean, well built, and apparently prosperous. It depends almost solely upon the agriculture of the neighbourhood.

The structure of the narrow valley along which I came to-day, and at either end of which, but especially at the eastern end, so much fertile land is to be seen, is very simple, but very interesting. Two ridges of elevated land, called respectively the North and South Mountains, run nearly parallel to each other from Windsor to beyond Annapolis and Digby, a distance of upwards of a hundred miles. The northern ridge consists of trap, resting upon a red sandstone, and forms the southern boundary of the Bay of Fundy. The southern ridge, called the South Mountains, consists of granite and of, more or less, metamorphic (Silurian and Cambrian) slates. The surface of the former has been crumbled by the action of the weather sufficiently to form over the greater part of the North Mountains a considerable depth of soil, which, like that of so many other trap rocks, is said by Dr Gesner to be rich and fertile. The granites and slates of the South Mountains have in general been slowly acted upon by the weather, and have unwillingly produced poor and scanty soils.

Between these ridges runs a long valley, widening towards the Bay of Minas, and affording at that extremity a larger expansion for the fertile alluvials of Cornwallis and Horton. In this valley lies, or formerly lay, a red sandstone deposit—that which still dips beneath the trap of the North Mountains—resting probably on some of the softer slates of the Silurian age.

In the drift period, when the whole of this country was submerged, the northern current, of which we have so many traces in these countries, rushing between the two lofty ridges of hard rock, scooped out the softer and less coherent red sandstones and marls and softer slates, and produced the existing valley, which, like the Bay of Fundy —a wider and longer excavation—has a north-east and south-westerly course.

And now, when the land was elevated to the existing level, the tides began to act as at present upon the Bay of Fundy, and to run round either end of the North Mountains, which, from Cape Blomedon to the Digby Gut, formed a long narrow island, having the Bay of Fundy on one side and the Strait of Annapolis on the other.

But the natural entrance of the tide into the strait

between the two ridges was through the Gut of Digby or Annapolis—a gut or opening through the lower end of the North Mountains into the Bay of Fundy—and here it would therefore enter when the waters reached it on their way up the Bay of Fundy. But through this narrow gut the tide could not advance with a velocity equal to that with which it ascended the open bay, and thus the tidal waters would round Cape Blomedon into the Bay of Minas, and, rushing westward towards Cornwallis, would meet the smaller arm of the tide which had come through the gut somewhere in the strait. Here a struggle would ensue, which would be repeated every tide, would shift its locality a little with the height of the tidal waters, and with the direction of the wind, but the effect of which would be to sweep into, and deposit on the site of the struggle, all the loose materials which the rains and streams brought down from either mountain-side, or which the tides themselves might tear from them. Thus a growing sandbank, and finally a bar, would be established in the strait, which would be a virtual water-shed, separating, as now, the tidal waters of the Bay of Annapolis from those of the Bay of Minas. On either side of this dividing line, the muddy waters of each bay would begin to deposit the rich slime which has consolidated into the fertile dyked land. And as the tendency always is, where such deposits take place, to raise the land highest near the water, the first formed dividing bank would remain at a lower level than the alluvial soil of newer formation, and thus a lake would be formed upon it, to dry up sooner or later into a bog or marsh. The great Carriboo bog, which still forms the water-shed and the origin of both rivers, stands on the site of the original bank, the scene of the once daily struggle of the two opposing tides.

The rest is easy. The deposits from the muddy water have gone on as they are doing now, till they have filled the whole of the space which the valley now occupies. And if the Annapolis dyked lands are less rich than those of Cornwallis, it is because the waters of the Bay of Fundy, coming in from the Atlantic, are less loaded with enriching matter as they enter the Gut of Digby than they are after they round Cape Blomedon; and because the discharge of fresh water into the west end of the valley is less, and the streams come through geological formations that yield their materials less largely to the waters which pass over them.

Annapolis is a quiet clean town, with considerable shipping capabilities, but little traffic. The drought, the potato failure, and other causes, had made the farmers poor; the home trade was therefore dull, and the good people of Annapolis in consequence discontented. As they could not think the cause of their interrupted prosperity was in any way to be traced to themselves, they were inclined to believe, with the Canadians, that it must be the fault of the Home Government, and that the certain cure was to shake themselves free of the mother country. I had not had much time to become initiated in local politics, but I was certainly pleased in listening to some of the warmer Annapolis politicians, to find them so very unsuccessful in making for this province anything approaching to a reasonable grievance against the Colonial Office. I pictured to myself Upper Canada in the character of one London *jarvie* saying to Nova Scotia in the guise of another, "What, no raw?" and thus exciting the ambition of his brother chip to discover or establish one. My present impression of Nova Scotia is, that it has a fine future before it. The friends of humanity will regret if its local rulers—its inhabitants, that is—shall suffer microscopic or imaginary evils to retard the discovery and development of its many natural resources, on which the rapid and sure realisation of that fine future so much depends.

On my arrival at Annapolis, I found that the steamer

to St John did not sail till Monday; so that I had two days to amuse myself in this neighbourhood. Part of one of these I spent in crossing the bay, and climbing the North Mountains, to visit a spot where I had been told that ice was to be met with all the year round. The day was hot, and the hill steep, and when we were fairly in the woods, I occasionally, for a short cut, forsook my guide and the trail, and fell among windfalls, so that I was not a little pleased when he announced our arrival at the spot. A windfall, in the English sense, usually means a bit of good luck; but when an Englishman gets into an American forest, he will soon unlearn this home sense of the term, and come to class it among unlucky events, with the occurrence of an alder swamp or a Carriboo bog.

The spot we had come to was a kind of notch in the side and summit of the mountain, where angular fragments and rocky masses of trap were piled one upon another, a little runner flowing down the centre of the notch. The whole was overgrown with mixed timber, chiefly hardwood, the roots of the trees fixing themselves wherever a holding-place among the stones was to be found. At various spots a freezing cold air was felt to issue from among the stones; and, on digging under the fallen leaves among the stony crevices, we succeeded in obtaining some lumps of ice, which, with the water of the brook and a little brandy—a prohibited drink in these parts—formed a refreshing beverage after our fatiguing ascent. This locality resembles those which have been described in different parts of Europe, where ice occurs, even in hot weather, among masses of collected rocky fragments. The air proceeds most probably from caverns in the mountain, which are filled with ice during the long and severe winters of this latitude, and are rarely melted by the warm air that enters them even during a hot and protracted summer.

I heard many complaints of the excessive drought in this part of the province. Parties who are badly off for hay are in the habit, in ordinary years, of sending to those who have hay to spare, three cattle at the beginning of winter, to receive back two in spring. This year five were already spoken of to get back three, and higher payments might become necessary.

The Bay of Annapolis is about twenty miles long, and at the foot of it stands the town of Digby. Several rivers flow into it from the South Mountains, among which the Moose river is distinguished by the occurrence of deposits of iron ore a few miles above its mouth. Another deposit of the same ore occurs on the Nictau river, which descends from the same mountains into the valley, about half-way between Cornwallis and Annapolis. Both ores are very rich, and that of Nictau abounds in casts of Silurian fossils.

Some years ago a company was formed for the purpose of mining and smelting these ores; and buildings were erected at the mouth of Bear River, where the manufacture was established and carried on. But differences arose among the partners, and the works were stopped. The site of the works is ten or twelve miles below Annapolis; and I was indebted to the kindness of Dr Leslie—a Scotchman possessed of the *perfervidum ingenium* of his country both in heart and head—for driving me to the spot. The site appeared to be well chosen, especially for convenience of shipment. There were also heaps of ore, and many tons of unfinished blooms, lying in the crumbling buildings, showing how summarily operations had been stopped. The furnaces and workshops were already falling to ruin, for want of that *stitch in time* with which masons, as well as tailors, can keep things in repair at a small expense. The locality is admirably adapted for the supply of iron to the markets of the two provinces, and of the Atlantic States; and if the adjoining forests yield fuel abundantly, and at a cheap rate, a prudently managed manufactory of malleable iron ought here to succeed.

I could not help sympathising with my friend the Doctor, when he discoursed of the extreme healthiness of the Annapolis district. Though he is the only medical man for sixteen miles one way and fifteen another, a fortnight will often elapse without a single summons. Were it not that the population increases, and that bones break sometimes, medicine and surgery might be banished the country.

The Nova Scotians have the reputation of being superlatively handy. "What will I do now?" issues from the mouth of a despairing Irishman; but with the emergency the resource not only springs up in the head, but actually rushes to the hands, of the Nova Scotian.

A farmer on the South Mountains will cut down lumber on his farm, and will convey it with his own horses to the shores of the bay. With or without the aid of a carpenter, he will lay down the lines of a ship. He will build it himself, with the help of his sons; he will even do the smith's work with his own hands. He will mortgage his farm to buy the materials, and will rig it himself. He will then load it with firewood from his own farm, and himself sail the ship to Boston, and sell cargo or ship, or both; or he will take a freight thence to the West Indies, if he can get it, and return in due time to pay off his encumbrances—or to sell his farm, if he have been unsuccessful, and begin the world anew. If the world were really to make up its mind to hang those who have no shifts, a vast number of our Irish fellow-subjects would be the first to taste the cord. The last survivor would be a Nova Scotian, unless, indeed, it were his fate to be strangled by my friend and subsequent fellow-traveller, Mr Brown of New Brunswick, of whose shiftiness I shall have occasion to speak in the sequel.

On the Sunday I attended service in the Episcopal church, and heard a sermon preached with a nasal twang so perfect that I guessed the preacher must be a Yankee. I was afterwards mortified to learn that he was a native of St John in New Brunswick; but I can honestly say for New England, that neither in the pulpit nor out of it did I meet, during my subsequent stay in the States, with any one so handy at speaking through his nose as this unhappy preacher of Annapolis.

The readers of Sam Slick naturally expect to hear many provincial expressions when they come to Nova Scotia. I was on the look-out for them; but whether it was that I did not fall in with any of the real blue-noses, or that the Queen's English is really better used than I had been led to expect, I scarcely heard a single peculiarity of expression during my stay in the province. Occasional guessings there were as to things which the guesser knew perfectly well—as when a man guessed his own age or his daughter's to be so-and-so, and the not unfrequent use of "admire" instead of "wonder at;" but what are these compared with our county provincialisms?

On Monday morning, the 13th of August, I embarked in the steamer for St John in New Brunswick. The weather was fine till we passed through the Digby Gut, and were fairly into the Bay of Fundy. A cross sea tossed us a little at the mouth of the gut, and by-and-by the fogs, and finally the rains and gusts of this bay, assailed us. The steamer was a poor affair, and among other freight had some sheep on board, for which the farmers of the Cornwallis and Annapolis districts find a ready market at St John. The breadth of the passage is about forty-five miles, which we accomplished by four in the afternoon; when I landed at St John, and took up my quarters in the hotel.

CHAPTER II.

Area and population of New Brunswick.—The lumber-trade, its benefits and evils.—It retarded and discouraged farming.—Emigration caused by a crisis in this trade.—City of St John.—Diminution in its import trade and in the provincial revenue.—Apprehensions as to the ability of the province to sustain its population.—River St John.—Rich river flats.—Average produce of Queen's and Sunbury counties.—City of Fredericton.—Farm on the St John.—Intervale land, its different qualities and values.—Emigration fever.—Indian corn as a fodder crop in England.—Opinion as to farming with paid labour.—Woodstock.—Quality and value of land in its neighbourhood.—Exhausting culture of first settlers. — Farming on Shares. — Charivari of the Mickeys of Woodstock.—Farm at Jacksontown. — Speculators in land.—Iron ore and iron smelting.—Itinerant lecturers.—Mouths of the Tobique and Aroostook rivers.—Potato breakfasts and meals in common.—Sowing of winter wheat on newly cleared land only.—Rust and wheat fly, remedy for.—Mellicete Indians on the Tobique.—Irish settlement and thriving settlers. — Healthiness of the province.—Grand falls and town of Colebrook.

BEFORE my departure from England, I had been invited by the Governor and House of Assembly of New Brunswick to visit that province, with the view of drawing up a report, to be presented to his Excellency and the Legislature, in reference to its agricultural capabilities. I had undertaken this task without very clearly understanding the nature of the duty, or of the country, and in the hope that it would not seriously interfere with my other plans in visiting the American continent. On my arrival, however, I very soon found that the extent of the province, and the slow rate of travelling, would compel me to devote some months longer to the work than I had originally anticipated; and, in order to complete it, I was subsequently compelled to delay, to a future opportunity, my intended visit to the more southern and westernly portions of the American Union.

The commercial, and I may say the entire internal and social condition of the province of New Brunswick, is in a transition state; and as all transitions occasion embarrassment and distress more or less general, wherever they occur, it has been the fate of this province to suffer a temporary check in its progress, in consequence of this transitionary state of things.

New Brunswick contains an area of eighteen millions of acres, of which about five millions are at present unfit for agricultural purposes. Its population is estimated at two hundred and ten thousand. With twice the geographical extent of the province of Nova Scotia, it has still a population about one-third less. It is therefore in a considerable less advanced condition than the latter province. Indeed, it was not till 1784 that it was separated from Nova Scotia, and formed into a distinct government.

The earliest inland trade of these northern provinces was confined in a great measure to the purchase, by way of barter, of the furs of wild animals collected by the native Indians in their hunting excursions. Next, and as settlers increased, the timber, or lumber trade as it is called, sprang up, and an apparently inexhaustible article of export was drawn from the boundless forests which stretched uninterruptedly over the entire surface of the province. The cutting of the trees, and the haulage and floating of them down the rivers, gave healthy employment to many men; the raising food for these men called agricultural industry into play; the export of the timber employed shipping, and afforded the means of paying for the British manufactures and West India produce imported in return; while the profits of the merchants erected towns and public buildings, improved harbours and internal communications, tempted foreign capital into the province, and generally sustained and carried it forward to its actual condition.

But, like other branches of industry, the lumber trade has always had its periods of activity and depression. When the demand was brisk and prices good, the trade was pushed eagerly forward; lumberers went into the woods by droves, and timber was shipped to England in quantities which over-loaded the market. Prices in consequence fell—those who were obliged to realise were compelled to sacrifice capital as well as profit; and thus mercantile crises, and many failures, periodically occurred among the colonial merchants. It was the over-trading of our own manufacturers in another form. The merchants of St John and the other lumbering ports were subject to these vicissitudes, not from any interference of home regulations, but through excessive individual competition among themselves. Still, on the whole the colonies gained, though many individuals were constantly suffering. And if home capital was lost to those who embarked it, it was a gain to the colony, inasmuch as it had been expended in paying for colonial labour, by which, directly or indirectly, colonial land had been cleared and prepared for the plough.

But such an export trade in the large could only be temporary. Land cleared of timber does not soon cover itself again with a new growth of merchantable trees. Every year carried the scene of the woodmen's labours farther up the main rivers, and into more remote creeks and tributaries, adding to the labour of procuring and to the cost of the logs when brought to the place of shipment. Hence, prices must rise at home, or profits must decline in the colony, and the trade gradually lessen. All these had already taken place to a certain extent, when the further increase of home prices was rendered almost impossible by the equalisation of the timber duties. In this alteration of our British laws, a large number of those engaged in the timber trade have been inclined to see the sole cause of the comparatively unprosperous circumstances in which they have recently been placed.

In so far as I have myself been able to ascertain the facts of the case, I think, with many patriotic colonists, that the welfare of these North American provinces would on the whole, and in the long run, have been promoted by a less lavish cutting and exportation of the noble ship-timber which their woods formerly contained, and which has already become so scarce and dear. Home bounties have tempted them to cut down within a few years, and sell at a comparatively low price, what might for many years have afforded a handsome annual revenue, as well as an inexhaustible supply of material for the once flourishing colonial dockyard.

At the same time, it is useless to lament over past mismanagement. It is easier to discern evils and their causes, after they have occurred, than to prevent even their recurrence. The cream of the timber trade being fairly skimmed off, the question, on my arrival in the colony, had assumed the matter-of-fact form—"How are we colonists in future to make our butter?"

It was an acknowledged evil of the lumber trade, that, so long as it was the leading industry of the province of New Brunswick, it overshadowed and lowered the social condition of every other. The lumberer, fond as the Indian of the free air and untrammelled life of the woods, receiving high wages, living on the finest flour, and enjoying long seasons of holiday, looked down upon the slavish agricultural drudge who toiled the year long on his few acres of land, with little beyond his comfortable maintenance to show as the fruit of his yearly labour. The young and adventurous among the province-born men were tempted into what was considered a higher and more manly, as well as a more remunerative line of life; many of the hardiest of the emigrants, as they arrived, followed their example: and thus not only was the progress of farming discouraged and retarded, but a belief began to prevail that the colony was unfitted for agricultural pursuits. The occasional large sums of money made by it induced also vast numbers of the farmers themselves to engage in lumbering—as a lucky hit in a mining country makes many miners—gradually to involve themselves in debts, and to tie up their farms by mortgages to the merchants who furnished the supplies which their life in the woods required. Thus not only were large numbers of the young men demoralised by their habits in the woods, trained to extravagant habits, and rendered unfit for steady agricultural labour, but very many of the actual owners of farms had become involved in overwhelming pecuniary difficulties, when the crisis of the lumber trade arrived, and stopped all further credit.

What added to the apprehension of the colonists at this time was the comparatively extensive emigration which began to take place when the demand for timber became less, and, consequently, for labourers to procure it. Undisposed to continuous farm-work, the lumberer left the province—as our navigators wander from country to country—to seek employment in Maine or elsewhere towards the West, where their peculiar employment was to be obtained. Even the pine forests of Georgia were not too distant for their love of free adventure. Unable to shake off their encumbrances at home, many of the embarrassed owners of farms also hastened to leave them—some in the hands of their creditors, without even the form of a sale—and made for the new states of the West, under the idea that in a new sphere they would be free men again, and that probably a less degree of prudence or industry might there secure them the competence which their own neighbourhood had denied them. No love of home, or attachment to the paternal acres, restrained either class of men; for these Old World feelings or notions have scarcely yet found a place among the Anglo-Saxons of any part of North America.

That such native-born and old settlers were leaving the province in considerable numbers, was construed into an indication that the province was inferior, as a place of residence, to the states and provinces to which they emigrated. Alarmists made it a topic of melancholy lamentation and gloomy forebodings; and, as in similar cases at home, party feelings laid hold of the emigration as a demonstration of the correctness of special party views, and exaggerated its evil effects. The departure of the working lumberers was a necessary consequence of the cessation of their favourite employment; and it was not considered that the moral character and habits of these men as a body, and the disheartened and embarrassed condition of the owners of the encumbered farms, rendered the departure of neither class a real loss to the population of the province; that the departure of both, in fact, was necessary, in order that the social state might have a fair chance of returning to a healthy, cheerful, energetic, and prosperous condition.

But if lumber, as a staple export, was to be insufficient to supply the future wants of the colony, in the way of paying for the necessary imports of West India produce and of flour, upon what were the colonists to fall back? Were the hitherto undervalued agricultural resources of the colony greater than they had been supposed? Could these 18,000,000 of acres really be made to support a population of 210,000 inhabitants, and thus enable them to dispense at least with the large importation of bread stuffs for which they had hitherto been yearly indebted to the United States, to Prince Edward's Island, and to

Canada? Or were the mines of the country of such value as to make up for the failure both of lumber and of corn, and to enable New Brunswick to keep pace in future progress with the adjoining states and provinces?

Such were the ideas and questions which had been passing through men's minds when I was honoured with the request to visit the colony, and give an opinion upon its agricultural capabilities. I trust that the result of my tour has been to inspire new hopes and awaken new confidence in the food-producing and population-sustaining powers of the land of this valuable colony, though it has lessened very much in my mind the opinion I had previously derived from books as to the extent of its mineral resources.

The city of St John is situated at the mouth of the river of the same name, which falls into the Bay of Fundy. It has a safe, though not extensive harbour, the entrance of which is defended by Partridge and other small islands. The principal part of the town is situated upon a rocky peninsula, which stretches into the harbour, but it is now extending itself in various directions over the adjoining crags and hollows. Notwithstanding the depression of trade which had for some time prevailed, the surface of naked rocks was, at the time of my visit, selling at the rate of £100 an acre for building purposes; and tasteful cottages, on picturesque sites, were springing up in the neighbourhood of the city. The older inhabitants of the city, the descendants of American loyalists, have many interesting facts to relate regarding its growth, upon what, sixty years ago, was a rocky headland, skirted by cedar swamps; and, considering the still generally uncleared condition of the province, and the position of the city itself, its progress has been at least as rapid as that of any of the greater cities on the Atlantic border of the North American continent.

Yet that there has been a serious change for the worse

in the trade of this part of the province is shown by the official returns of exports and imports from the port of St John, for the three years ending in December 1848. These are as follows:—

	1846.	1847.	1848.	Diminution in 1848.
Imports, .	£977,683	£1,070,514	£588,422	£482,092
Exports, .	810,742	632,612	588,466	44,146

Thus the exports have been regularly diminishing during these years, and consequently, though not immediately, the imports also. And, as affecting the trade with the mother country, it is an important fact that, of the total decrease in 1848, compared with 1847, no less than £336,100 were in the imports from Great Britain. Of this sum the diminution in the importation of—

Manufactures of cotton, woollen, linen, and silk, was	£157,421
Iron, wrought and unwrought, ...	46,267
Copper, ...	9,319
Hardware, ...	22,951
Leather manufactures, ...	1,923
Cordage, twine, and canvass, ...	47,044
Tea, ...	6,975

Thus, all our home industrial interests are concerned in the prosperity of our colonial possessions, and we help our own pockets when we contribute to their material advancement.

Another way in which this falling off in the exports and imports of St John had affected, not only the city, but the province in general, and had made people fretful and uneasy, besides embarrassing the Government, was the great reduction it caused in the revenue, a large portion of which is derived from the duties levied at the custom-houses, and from a small export-duty on timber. Thus, in the three years I have mentioned—

	1846.	1847.	1848.
The total revenue was .	£127,336	£127,410	£86,437
The revenue from customs,	30,961	31,912	2,711
The export-duty on timber,	22,664	16,553	18,252

It was natural, therefore, that all parties should feel uneasy at such a state of things—a falling off in the revenue of nearly one-third—and I was not surprised to hear charges of the gravest nature occasionally made against the competency, and even the honesty, of the existing provincial Government; or the Canadian grumblings re-echoed, that connection with England, after all, was the main source of colonial sufferings. It is human nature, and especially the nature of political parties, to ascribe to neglect or unskilfulness on the part of man what physical or moral laws render it impossible to prevent.

I shall have occasion hereafter to speak of the extensive diminution of the wheat crops in North America; but I may here merely mention, in connection with the other causes of colonial depression, that the united failures, for a succession of years, of the wheat and potato crops were further just causes of disquietude to the provincial population. It must have alarmed those who were not themselves possessed of agricultural skill, or who had not had an opportunity of looking at the whole province with an agricultural eye, to learn from the published returns that, in 1847, wheat and flour, to the amount of about 240,000 bushels, were imported into New Brunswick, and that the estimated value of all the bread stuffs imported during that year was £280,000 currency.

Reckoning all the grain imported at the average high price of 40s. a quarter, this sum would imply, that at least 140,000 quarters of grain, or their equivalent in flour, were imported in 1847—a quantity sufficient to feed at least one-half of the whole population of the province. It was natural, therefore, to say—if the lumber-trade fail, and we can raise at home only enough of food to support one-half of our population, where are the means to be obtained by which the other half is to be kept alive? In such circumstances of doubtful anxiety, the political condition of the province must, on the whole, have been satisfactory to have given rise to the very small measure of excitement which it was my fortune to meet with during nearly four months that I spent in the province.

The river St John empties itself into the harbour through a narrow passage between high opposing cliffs of metamorphic-slate and limestone rocks. At low tide, a long rapid and a considerable fall exists at the mouth of the river; but the tide rises twenty-six feet, which is sufficient to equalise the level of the outer and inner waters, so as, for a brief space before and after high water, to allow vessels of considerable tonnage to ascend and descend with safety. Well-appointed steamers ply upon the river between St John and Fredericton, the seat of Government—a distance by land of sixty-five miles, and by water, I believe, of about ninety. In spring and autumn, when the water is deep, they ascend to Woodstock, which is sixty-two miles higher; and when the contemplated improvements are made in the river, small steamers are expected to mount as high as the grand falls, which are seventy-three miles above Woodstock. This extensive natural inland navigation—nearly equal in length to that possessed by the state of New York—will every year become more valuable to the colony.

14th August.—At one P.M. I embarked on board the steamer for Fredericton, where I arrived at 8½ P.M., being at the rate of about twelve miles an hour. The day was fine, and the sail very beautiful. For the first thirty miles the river is wide, and has rocky banks of varying height and form, covered with a natural forest growth, except where the hand of man has been busy in partially clearing and establishing farms. The rocks, at the outset, consisted of mixed limestone and slate, then, for a considerable distance, of trap and metamorphic slates, as far as the head of what is called the Long Reach. Then turning us sharply to the left, and narrowing the river for a few miles, a ridge of granite, visible only on one side of the stream, succeeded to the trap; after passing which we emerged into an open and flatter region, over which grey sandstones, of the coal formation, extended and accompanied us all the way to Fredericton.

The trap country reminded me of some of the thinly-peopled districts on our Highland lakes. It was covered in many places with a sandy drift, and bore, in general, a mixture of broad and narrow leaved trees. On the granite, broad-leaved or hard wood prevailed, the poorest soils bearing only the white birch. Endless pine forests covered the sandstone soils, where drift from other formations, or the sorting action of flowing water, had not modified their natural character.

Through the first twenty miles of this sandstone formation extends a very beautiful portion of the river. From the north-east enters the Washademoak River; and fifteen or twenty miles above its mouth, the Salmon River, after traversing the Grand Lake, escapes into the St John. At the mouths of both these rivers, the St John widens, and is studded with several large and fertile islands; while the low *intervale* land, as it is called, stretches sometimes a couple of miles from its banks. Gagetown and Scovell's Point, on its opposite shores, are centres of rich land, which appeared to be tolerably well farmed.

Many emigrants, with money to purchase farms at two to four pounds an acre, might settle comfortably here. This alluvial land has been long famed for its grass and its produce of hay. In this country, where hay has hitherto been the chief reliance for the winter food of stock, the produce in hay is generally considered a test of the value of a farm, either to rent or to buy. In renting land, not a very frequent practice, a pound of rent for each ton of natural hay produced by the farm is, on the St John, considered a fair equivalent. The produce in grain is not taken into account. Hay sells, according to the season and locality, at 35s. to 50s. a ton.

These low lands are liable to be flooded when the ice melts in spring, but they are, nevertheless, very healthy. There are no agues in the country! I have heard of none, indeed, in the whole province, even where waters and bogs and marshes most abounded. These spring floods, no doubt, contributed to the richness of the land; but the best situated or most esteemed farms here are those which consist partly of this low intervale and partly of upland.

The soils in general are light and loamy, as we should expect in a sandstone country; and, therefore, adapted to the culture of Indian corn, which in this part of the province has been considerably extended during the last seven years—I suppose since the wheat crop became less certain. From the mouth of the Washademoak river, in ascending to within a dozen miles of Fredericton, the St John carries us through the centre first of Queen's, and afterwards of Sunbury county. Much of these counties is still in native forest; but the general productiveness of the cultivated land, and something of the husbandry and cultivation, may be judged of from the following returns as to the maximum, minimum, and average produce, in imperial bushels, of the crops usually cultivated in these two counties.

	Queen's.			Sunbury.		
	Max.	Min.	Average.	Max.	Min.	Average.
Wheat, .	20	8	14½	30	12½	19¼
Barley, .	18	18	18	40	20	30
Oats, .	60	13	29	50	30	38½
Buckwheat,	50	15	27½	60	20	33¾
Maize, .	50	20	33½	80	35	51¼
Potatoes, .	400	100	181	400	100	204
Turnips .	1000	200	550	800	200	500
Hay, .	3 tons	1 ton	1½ tons	3 tons	1 ton	1½ tons

The produce of the potato in this table is small, because of the failure of this crop during the last few years. The turnip culture is not general as yet, but is extending. The intervales of Sunbury county are especially productive in Indian corn.

I have seldom seen anything of its kind, which, as the sun declined, seemed to me more beautiful than the banks of the St John in this county, as we passed Majorville and Sheffield, and approached the mouth of the Oromucto river. The river, full to the lip, reflecting the light of the western sun towards which we were steaming, shaded on either bank by rows of the American elm—which I here saw in its great beauty for the first time, and which, every time I have since seen it growing wild in its favourite localities, has always struck me as the loveliest of American trees—and beyond the banks broad fields of Indian corn in the full rich green of its still unripe growth. In this there was newness enough, perhaps, to give it a charm to my eye, which would not have been seen by one more familiar with the country; but, after making a large deduction for this, there remained beauty enough over to make this part of the river, at this season, interesting to the oldest dweller in the province. I have since seen no river scenery in America which has left on my mind a livelier impression than this part of my voyage on the St John.

Fredericton is the seat of Government. It stands on a flat of level intervale land, in some places nearly a mile in width, and raised about thirty feet above the river. Upon this level, thirty years ago, there were only two or three houses, surrounded by thickets and cedar swamps. It is now a considerable town with five or six churches, besides a cathedral, built under the auspices and by the exertions of the present bishop. It has a University, (King's College,) a dissenting academy, a grammar school, normal school, court houses, Government offices, legislative halls, well-built streets, barracks for a thousand men, and a population probably of four or five thousand people. The soil of the level on which it stands is light and sandy, resting at a variable depth on a bed of clay. The hill-slope behind is in general very stony, and costly to reclaim, and is covered for the most part with the native forest of pine. Opposite the town is the mouth of the Nashwauk, a considerable stream, which here falls into the St John; and a little above the town that of the Nashwauksis, or little Nashwauk. The former is navigable for some distance into the interior.

The St John itself is here confined within higher sloping banks, and is about three-quarters of a mile wide. The influence of the tide is observed about four miles above the town; and at Fredericton it seldom rises more than fifteen inches, so that it may be said to be situated at the head of tide-water. Steam and horse ferries are established on the river, by which a regular communication is kept up with the opposite shore.

16*th August.*—At Fredericton I was joined by Mr James Brown, a member of the Provincial Assembly, and by Dr Robb, Professor of Natural History in King's College, who accompanied me during the whole of my subsequent tour in the province, and to both of whom I was indebted for much information and assistance. The familiarity of the former with the practical agriculture and economical condition of the province, and of the latter with its geology, in so far as it had previously been made out, enabled me to arrive much more rapidly at satisfactory conclusions, in regard to the agricultural capabilities of the province, than I should otherwise have been able to do.

Early this morning we started in an open carriage up the right bank of the river, and stopped to breakfast at Oakhill, a farm lately bought by Mr Jardine, a merchant of St John, and occupied by Mr Gray, a Scottish farmer, who had recently quitted the neighbourhood of Girvan in Ayrshire, for the purpose of settling in New Brunswick. We found him busy improving and enlarging his farm-buildings, and after breakfast we walked over his farm. As it is the first farm I examined in the province, I may be permitted to give some general description of it.

It consists of a thousand acres in all, of which two hundred are cleared, and eight hundred in forest, chiefly soft (pine), but some of it hardwood. It contains land of three kinds. *First*, an island in the river of eighty acres, to which I crossed, and found it a free grey loamy clay full of natural richness, and subject to be overflowed only twice during the last thirty years. *Second*, intervale land, generally light and sandy, but bearing in some places good turnips, and resting upon a loamy clay resembling that of the island, at a depth in some places of no more than eighteen inches from the surface. I do not know the extent of this intervale, on which the house stands. *Third*, the rest is upland, on the slopes generally very stony, but in other parts of the farm capable of being easily cleared. But two hundred acres of cleared land form a large farm where labour is scarce and dear.

This farm cost about two thousand pounds currency (£1600 sterling), or two pounds an acre over head; and this may be considered about the present price of such mixed farms on the upper St John. It had been exhausted by the last holder by a system of selling off everything—hay, corn, potatoes—the common system, in fact, of North America of selling everything for which a market can be got; and taking no trouble to put anything into the soil in return.

Farming on shares, the Metayer system, is practised in the Provinces and New England states, more than our home method of paying rents. In this way a man who has nothing receives a farm, with stock, implements, and seed, from the owner, provides all the labour or works the farm, and receives half the produce of cheese, stock, grain, potatoes, &c. This is said to be, in general, rather a better thing for the cultivator than for the owner. In most cases, however, there are specialties in the bargain, the owner receiving more or less according to the condition, position, or richness of the farm. I have already spoken of the system of reckoning the value of land for renting by the quantity of hay it will produce.

Leaving Mr Gray's, we continued our drive up the river. Hitherto we had been upon the grey sandstones, some beds of which, from the quantity of earthy felspar cement they contain, are capable of yielding soils of fair quality. We now came upon the slate rocks, and upon these we continued, with the intervention of a narrow band of red sandstone, and occasional masses of trap, or trap-like metamorphic slates, for upwards of twenty miles. We then crossed a broad zone of granite, which, like a long ribbon, stretches across the province in a north-east and south-west direction, from the Bay de Chaleurs down to this part of the River St John, and hence over into Maine.

On the slates good land often occurs; but, as the river banks are high, a journey along the river side is not favourable to an estimate of the quality of the upland. The granite region, and much of the slate country adjoining it, are thickly strewed with stones; though the soil itself, as seen among the stones, or where the stones are removed, is very good. Rich intervale land and occasional islands were seen along the river and the cleared openings we passed. The frequent boldness and beauty of the landscape, the varying forms and fresh verdure of the trees—elm, butter-nut, black-birch, maple, oak, beech, cypress, and numerous pines—with the good roads along which we passed, and a good dinner by the way, and agreeable companions, full of information new to me, made the day glide on very pleasantly, till we reached the mouth of Eel River, a distance of fifty miles from Fredericton, where we took up our quarters for the night.

Of the intervale land there are three varieties at least along the river St John. The best is that which is just above the present high water, or usual flood level, of the river. It is generally a free rich loam, easily tilled, and producing large returns of hay, a crop here so highly valued. The next is a ledge from eight to twenty feet above the former, which is usually of a lighter quality, and less valuable—sometimes sandy, gravelly, and almost worthless. On these dry worthless sands, and as a token of their worthlessness, springs up the fragrant everlasting, *Gnaphalium polycephalum*, with which I had the opportunity of becoming very familiar before I quitted the province of New Brunswick.

At a higher level still, the third intervale land occurs; and besides the sand and gravel of which it not unfrequently consists, it carries stones or boulders, occasionally in considerable numbers.

These different intervales are in reality successive terraces, rising to different elevations above the existing bed of the river, but showing the different heights at which the water has stood since the stream began to flow in its present channel.

I have alluded in the commencement of this Chapter to the emigration from the province, which to some had been the cause of much anxiety. I heard at this place of the first striking example of the height to which the emigration fever will run. About eight miles from the mouth of the Eel river lies the Howard settlement, situated on a tract of good second-rate upland, in the township of Dumfries. In this settlement a farm is at present offered for sale, consisting of 200 acres, of which 60 acres are cleared. Four acres are in wheat, 2 in Indian corn, 24 bushels of oats have been sown, 1½ of buckwheat, and 20 of potatoes. There are also four cows, two oxen, two horses, two heifers, fifteen sheep, 20 tons of hay, with a house 20 feet by 30, and a barn 30 feet by 40. The whole offered for £140 currency (£112 sterling.) The only condition is that of ready money. The owner is said to be mad to go to Wisconsin. It ought not to surprise us that some of those who have shifted once—breaking loose from all ties of place and blood—should after a time have another access of the roving fit, and, right or wrong, insist on moving a second or a third time. Changing their country is to many like a change in their religion—they don't know when or where they ought to stop.

17*th August.*—This morning, the rest of our party proceeding by land, Dr Robb and myself went up the river in a canoe, as far as Woodstock, that we might see better the general character of the country on either side of the river, and look out for a bed of rock-salt, which a sharp New Englander alleged to exist somewhere by the way. As to the latter point, as the river runs all the way through old slate rocks, our exploration was of course unsuccessful; but we found a beautiful white vein of quartz, which looked white and glistening like salt, and had most probably been mistaken for it. The shallowness of the river at this season of the year made the pulling and polling heavy, so that we spent a large portion of the day in going over these twelve miles.

A few miles below Woodstock we stopped to look at a farm on the left bank of the river, owned and occupied by Mr Rankin. It consists of about 1100 acres of intervale and upland, of which 100 were in crop, meadow, and pasture, chiefly on the intervale land. It is an upper intervale, resting on accumulations of gravel and sand, and therefore for the most part light, and sometimes sandy. Wheat, oats, Indian corn, and potatoes, are the crops raised—the corn more largely since the failure of wheat and potatoes commenced. The wheat on the ground this year promises 25 bushels an acre, potatoes yield an average of 150 to 200 bushels. The Indian corn always ripens, yields about 50 bushels, and is at present the most profitable crop.

The straw of the Indian corn is a very valuable fodder. If cut before it is dead ripe, it is as valuable as hay, and the cattle eat it as readily. Of this fact I afterwards met with many corroborations, though, both in the Provinces and in the Northern States, the wasteful practice of leaving the straw in the field uncut extensively prevails. Besides the grain, as much as three tons of excellent fodder may be generally reaped from an acre of Indian corn of the taller varieties. The advantage of this, not only in saving food, but in manufacturing manure, every home farmer at least will understand.

Indian corn has at various times been recommended as a grain crop to our British farmers. But our summers are not dry and hot enough to make it certain as a grain crop. It is worthy of a trial, however, as an occasional fodder or green crop on our lighter barley soils. A well-manured field would raise a large crop of green stalks, which are very sweet, and it might be profitable either for soiling or for making into hay.

The stock kept by Mr Rankine was seventeen milk cows and thirteen other cattle, which consume on an average about 60 tons of hay. Butter and cheese meet with a ready sale. He had also sixty-five sheep, which average, including lambs, 6½ lb. of wool. This his family manufactures into excellent homespun checks and tartans, which are sold in the neighbourhood.

The reader will naturally inquire, as I did—"Here you possess a farm of 1100 acres, and you have only 100 cleared, and of this 100 only 50 in arable culture; why don't you clear more, and farm more extensively?" "We clean up two or three acres every year of the lumbered land (land from which the timber has been cut,) because it is unsightly, not because we want it—we have as much land already as it is profitable to cultivate."

And this I afterwards found to be a very prevailing opinion, not only in New Brunswick and the other Provinces, but in the United States, as far west as the foot of Lake Erie, the limits of my own tour. It is profitable to farm as much as can be cultivated with the members of a man's own family—it is not profitable to farm with paid labour. That such an opinion should be so widely entertained shows that it is the result of a very wide experience. At the same time it may only be true of the system of farming hitherto adopted by the parties who entertain it, or inculcate it upon others. It may not be true of another or more improved system.

In reference to the agricultural capabilities and improvement of the colony, and especially in reference to the question of its being desirable as a settlement for British farmers possessed of capital and skill, this question is a most important one. I shall briefly state the general result of my inquiries when I shall have gone over a larger portion of the province.

Woodstock, the chief town in the county of Carlton, is advantageously situated at the mouth of the Meduxnakik, on the right bank of the St John. It has four churches, a grammar school, and about 3000 inhabitants. It is likely to flourish, both because it is connected with one of the richest agricultural districts in the province, and because here the road to Houlton in Maine branches off, and it ought therefore to be the centre of the traffic with the upper portion of that state. The boundary line between Maine and New Brunswick runs about ten miles west of Woodstock.

From the mouth of Eel River, twelve miles below Woodstock, where we left the granite region, the soil has gradually improved; and from the neighbourhood of the town northwards to the Grand Falls, and on both sides of the St John, it is generally equal in quality to the best upland in New Brunswick. The Cambrian appears in this region to have given place to the Silurian slates, and the soil resembles in some degree those of the upper Silurian slates, which I afterwards saw in the wheat region of western New York.

The president of the county Agricultural Society drove me a few miles inland to what is called Scotch Corner, in the direction of the Maine boundary. A long, flat, second terrace, or intervale, stretches inland about a mile from Woodstock. The cleared land on this flat is valued at £5 an acre. The country as we proceeded was beautifully undulated—chiefly covered, where the forest remained, with large hardwood trees. The rock maple and black birch, mixed with butter-nut and elm, indicate good, deep, heavy land—the beech a heavier soil.

At Scotch Corner, I saw a fine second crop of potatoes, grown without manure; and I examined a field of oats, which was the tenth grain crop (oats, pease, and buckwheat in succession) grown on it without manure. The soil consisted of fragments of a shivery slate, which crumbles readily, and which, at a depth of sixteen inches, rests on the rotten slate rock.

Old Country agriculturists, or those who, without being farmers themselves, condemn every practice which differs from what they have been in the habit of hearing commended at home, cannot fairly appreciate the circumstances of the occupier of new land in a country like this. For ten years—for eight, or twelve, or twenty years in other localities—this new land requires no manure to make it yield good crops. On the contrary, the addition of manure makes the grain or grass crops at first so rank that they fall over, or lodge, and are seriously injured. Thus, to a settler on new land, which he clears from the wilderness, manure is not only unnecessary, but it is a nuisance; and hence he not only neglects the preparation of it, but is anxious to rid himself in the easiest way of any that may be made about his house or barns.

Careless and improvident farming habits were no doubt thus introduced, so that, when at last the land became exhausted, the occupiers were ignorant of the means of renovating it. Old habits were to be overcome, new practices to be adopted, and a system of painstaking and care to which they had been previously unaccustomed. Hence, no doubt, the reason why I was almost everywhere told that it was cheaper and more profitable to clear and crop new land, than to renovate the old.

Still, because of these future evils, we are not justified in speaking contemptuously of present holders of new land, who, being desirous of making the most of it with the means at their command, waste none of their attention upon unnecessary manures. These men form that body of pioneers in American agriculture, who, having done their work in clearing and superficially exhausting one tract of land, move off westward to do the same with another, selling off each farm in succession to men possessed of more knowledge than themselves; whose skill and industry must bring back the fertility which had disappeared under the treatment of their predecessors, and who have no temptation to fall off into negligent modes of farming.

According to this view, the emigration of this class of wilderness-clearing and new land-exhausting farmers, is a kind of necessity in the rural progress of a new country. It is a thing to rejoice in rather than to regret, as I found some of my New Brunswick friends doing. At all events, I believe it has had a considerable influence in setting in motion and in maintaining that current of human *movers*, which, beginning in a tiny rivulet at Newfoundland, gathers, as it advances westward, till it forms the great river which is now flowing so fiercely into California.

On my return from Scotch Corner, I visited a fine farm belonging to my conductor, the president. It is let on shares to an English farmer. The landlord stocks it, the farm seeds itself, and the farmer does the farm work, and receives half the profits. The drought of the season had lightened the grain crops, but I saw some fields of excellent turnips, and some of oats averaging about twenty-six bushels an acre. In the yard was a fine herd of stock, chiefly mixed Herefords and Devons, with a little short-horn blood. They were coarse and thick in the skin, but probably on that account better adapted to the climate. On the whole, though the owner thought I did not sufficiently praise them, I did not afterwards see in the province a herd in all respects equal to them.

At Woodstock, in the evening, we were gratified with an interesting musical entertainment. It seems that the Orangemen are numerous in some parts of New Brunswick, and that Woodstock has its full share of them. Some twelve months or more ago, a riot took place between them and the Romanists, (Mickeys, as they are here called,) attended by the destruction of a considerable amount of property, which the county of course was called upon to pay. But the county applied to the provincial House of Assembly, to have the sum in whole or in part paid out of the provincial treasury; and in reference to this matter, my fellow-traveller Mr Brown, as a member of assembly, had given a vote which was unsatisfactory to the Woodstock Orangemen. Hearing of his arrival, therefore, instead of lynching him, as they might have done a little farther West, they serenaded us all at the hotel until near midnight with a charivari of all the most discordant noises, vocal and instrumental, which the tongs, kettles, saucepans, and throats of Woodstock could produce. There were also tar-barrels and bonfires on the occasion, and finally a burning in effigy. Fortunately the budding Orangemen did not personally know the man they thus delighted to honour; so that Mr Brown himself flitted about the blazing barrels, and enjoyed the burning fun as much as any of them.

18th Aug.—Though a little tired with the dissipation of the previous night, we started by half-past seven A.M., to proceed up the river as far as the Grand Falls.

On leaving the town we turned to the left, forsaking the river, and taking an inland road, for the purpose of passing through some of the new settlements in this county. Jacksontown, at the distance of five or six miles, was the first settlement we entered upon. It is about fifteen years since it was first commenced. The land is good, though now and then patches overspread with sandy drift occur, bearing the ill-omened Everlasting as their natural produce.

I stopped a few minutes at Hannah's farm, on which reapers were at work. It consisted of 200 acres, of which 80 were cleared. This, besides building a nice house, he had cleared with his own hands in thirteen years. The cradlers, who were cutting his grain, received from 1 to 1¼ dollars a day, besides their victuals. They were lumberers, who at this season of the year are usually at home.

Most of the land in this region is granted; and here I first began to hear from the mouths of working farmers the complaint which has been made successively in all the provinces, and is not unknown in the newer States of the Union, that large portions of the best land have been granted—that is, sold at the Government price—to speculators, who buy for the purpose of holding on till the neighbourhood is improved, and then selling at an increase of price. This is provoking to poor men who wish to buy farms, and settle near their friends; but it is injurious to the whole community, in a country where roads are comparatively few, and desirable lands in many localities are at present worthless, because miles of tangled forest shut them out from communication with the world.

It is very difficult either to remedy or to prevent this evil. The provincial Government are endeavouring to make it less frequent in future, by limiting the extent of individual grants, and by requiring that a certain proportion of each grant shall be cleared within an assigned number of years.

Notwithstanding the obstacle presented by the preemption of so much of the good land, in this neighbourhood, by persons who do not intend to improve it, the extension of this settlement has proceeded rapidly of late. The failure of the lumber-trade is inducing more young men to adopt what is, after all, a surer mode of living; and back lots are taken up and being cleared, where the line of farms next the road is already disposed of. The same is the case on the Maine side of the boundary line, where the land is also good and settlers fast pouring in.

Iron ore is abundant in this neighbourhood. It is of the hematite variety, and a smelting furnace has lately been erected within a short distance of Woodstock, for the purpose of smelting it. It is reduced by means of charcoal, and the hot-blast is employed. The iron obtained, up to the time of my leaving the province, was too brittle for casting, but it was said to make good malleable bars and steel. I visited the works on my return down the river, and it appeared to me, considering all the circumstances, that the company had begun their works on too large and expensive a scale. Some of the less ambitious establishments on the Housatonic river, in Connecticut, would have probably been safer models for them at first, than the huge smelting-furnaces of Scotland and Wales. The success of these works, however, is of great moment to the province, inasmuch as their failure would be a serious check to future adventures of capital in similar undertakings.

The land on this day's journey continued good nearly the whole way; and the crops of oats and potatoes were more like good crops in Scotland than anything we had yet seen. The English or Scottish farmer who may think of settling in this country must not consider himself as quite out of the world in these parts. There are wandering teachers, who supply with knowledge the thirsty cultivators in the humblest villages. Notices are stuck up in the inns, or are printed in the newspapers, or are spread in the form of handbills, such as two I met with to-day—" Mr Humphreys intends to lecture in this village, during the current week, on electricity and the electric telegraph."—" Mr Dow intends to lecture on physiology and anatomy during the present week; we hope our friends will give him full houses during his stay among them."

That these wanderers receive encouragement, not only here but on the other side of the border, is shown by an amusing circumstance told me subsequently by a fellow-traveller, when on my way, through Maine, from Bangor to Boston. Though a Bangor man, he had property and business which took him frequently into Georgia. " When on his way to Boston, on one occasion, with a friend, who had also been with him in Georgia, they dined at a hotel, where they saw opposite to them at table two New Englanders, whom they had last seen peddling in Georgia. 'Well,' says his friend to one of them, 'when did you quit your peddling in Georgia?' The questioned made no reply, but, swallowing his dinner expeditiously, as a New Englander can, he went out of the room, and, waiting for my friend and his companion, accosted them with, 'For heaven's sake, say nothing about the peddling. We have have been up to Maine, and, as our wares were out, we took to the lecturing. It's not a bad trade; we have made sixteen dollars a-day since we began. I take astronomy, and he does the phrenology. We have been lecturing in Bangor, and we have promised to go back. We had an invitation to go down to Bucksport, but we heard of some people there who knew quite as much as ourselves, so we declined. Now, you won't say anything about the peddling.' "

We had returned to the St John, dined on its banks at an inn, situated at the mouth of what is called Buttermilk Creek, and had driven nearly thirty miles further, when we found ourselves at the mouth of the Tobique, a river which comes from the east, and falls into the left side of the St John. This position is remarkable for an extensive second interval or terrace, of great extent, and of comparatively rich land, which is all cleared and settled, is finely cultivated and improved, and is pleasant to the European eye, from the number of well-built, clean, comfortable-looking houses, which are spread over the flat. The place has also its Episcopal church, and, on the whole, appeared to me rather an enviable locality, though at present a considerable distance from the world. It is opposite the mouth of the river Tobique, which flows through a still, wild, but agriculturally capable country, which fifty years hence will sustain a considerable population. This flat, therefore, is likely to be the site of a future town of some importance.

The upland here is also of good quality. A farm of 200 to 250 acres upon it, with 40 or 50 cleared, and a good house, will sell at present for about a pound currency an acre.

Three or four miles further of a pleasant drive brought us to the mouth of the Aroostook river, which flows from the interior of Maine, and empties itself into the right side of the St John. This is an important river, as, in seasons of high water, it admits of about 400 miles of inland and lake navigation, (*Gesner*,) and passes through a rich valley, forming one of the best farming regions of the State of Maine. The valley of the Aroostook was one of the most valuable portions of the disputed territory, and one which both New Brunswick and Maine desired to possess. At present, this valley forms a rich and almost untouched lumber country, from which large quantities of timber are floated down to the St John on the waters of its river. By treaty, the free navigation of the St John is secured to the produce of the Aroostook. We stopped for the night at an inn at the mouth of this river, which, in the height of the lumbering season, is alive with swarms of lumberers, whose hobnailed shoes had everywhere indented the wooden floors of the rooms and passages. A few scattering men were already on their way up to the woods.

Sunday, 19*th August*.—The English traveller, who starts on a North American tour, must shake off some of his home habits and notions. Potatoes to breakfast, which he will see everywhere—without which, I believe, in these provinces, a breakfast would be considered incomplete—is not an American custom solely, as I have met them many years ago in the west of Scotland, and, if not of Irish descent, is probably a home provincial custom, extended, by the necessity of circumstances, in the foreign provinces. A common table for all will at first surprise him more. The driver and his passengers, the hired and the hirer, and the humblest wayfarer who may desire to dine when your dinner is ready, sit down together. We had ordered our own meal to-day in our own sitting-room, but we found ourselves, after a time, seated side by side with ill-appointed lumberers, in fustian jackets, without any one, except myself, appearing even to feel that there was anything out of rule in such intrusion. We were close to the boundary of the country where all men are born free and *equal.*

The wheat crop in these northern parts of America has a history which is interesting, not merely to the practical agriculturist, but even to the political economist of the broadest views. I shall have occasion hereafter to return to this subject, in discussing the relation of the American wheat-producing capabilities to our home agricultural condition. I shall here, however, mention two particulars of a practical kind.

In the first clearing of a piece of woodland, when he hews his farm out of the forest, the new settler sows his wheat in the autumn. The winter snows fall and cover it, till one sweeping thaw comes in spring, when the green blades spring up under the influence of the sun, and ripen into a healthy crop. But after the woods have been cut back, and the land has been more widely cleared, the continued covering of snow is not so certain. Spring comes with partial thaws and freezings, which throw out the winter wheat, and kill it in whole or in part. The only practical remedy adopted for this is to sow spring wheat, which rushes up and ripens rapidly, but yields a grain which is said to be not equal in quality to the winter corn. This fact has an important bearing on the supply of first-quality flour to the American and European markets.

Again, in many localities the wheat crop is liable to rust, and in many more the wheat-fly has come like a pestilence, and almost put an end to the cultivation of it. The practical remedy for these two evils is to sow bearded wheat. Of this two varieties are here sown, both spring wheats. The one is known as the old bearded red, and the other as the Black Sea wheat—a white bearded variety. These are supposed to be less liable to the attack of both the vegetable and the animal plague; but even of these varieties, average crops, until the present, which is a very promising year for grain, have been by no means to be depended upon.

After breakfast, I drove back to the Tobique, and attended service in the Episcopal church. The service was well read, but the congregation was small, and the horses and waggons tied to the railing showed that most of the people had come from considerable distances. The Episcopal clergy of the province have hitherto, I believe, been almost entirely supported by remittances from the Propagation Society at home. These, as the country becomes settled, must, of course, be withdrawn; and the Bishop is, I understand, exerting himself very much in preparing the people for the coming emergency.

At the mouth of the Tobique, on a flat high intervale of good land, upon the opposite or left bank of the St John, is situated a native village, of twenty or thirty houses, containing about a hundred and twenty Indians. After forenoon service, I crossed to the village in a canoe, and was informed by my Indian ferryman that the population was nearly all collected in the chapel. I went towards it, and, as I approached, a few small children ran screaming from me in terror, and beat lustily for admission at the chapel door. On reaching the chapel, I was admitted by some of the older people attracted by the noise. I found a few well-dressed Indians, men and women, seated in pews, but the herd of the squaws and dirty children squatted on the floor at the end of the chapel. There might be thirty, young and old, in the place, and two men of the tribe, kneeling at a respectful distance from the altar, were doing their best to chaunt the service. I staid a few minutes, and then, having put a bit of silver into the collection-box, and distributed all the copper coins I had among the little ones as I went out, I left them apparently not dissatisfied with my intrusion.

On walking down the village, I met one or two of the Indian men, and on asking them why they were not at prayers, they said they did not belong to the priest's people ; but whether they were Protestants or heathens, I did not clearly make out. I then went into several of the cottages, and found in some only women and children, in others brawny men devouring wild-berries, which the women had been collecting in the woods. Some of the cottages were clean, and the inmates comfortably clad. This was especially the case with the house and family of the chief, whom I visited after his return from the chapel, where he had been officiating in the absence of the priest. He was an old man, small in size, but with a very intelligent face.

These Indians are of the Mellicete tribe, as I believe are all the Indians of New Brunswick. They are a robust race of men, not half civilised however, and never to be weaned from their love of the woods. At this place they own a reserve of 16,000 acres, a large portion of which is choice land, which they will never cultivate, and which must by-and-by be sold by Government in some way for their benefit. There are altogether in the province some 1400 Indians, and they hold reserves of about 63,000 acres of land.

I returned to the Aroostook to dinner, and afterwards went on to the Grand Falls. This was a drive of four hours, and, by the aid of the good roads, we reached the Falls—otherwise, the town of Colebrook—about 8 P.M. The roads in New Brunswick are really good, and very creditable to the province. This opinion, which I had already formed, was subsequently everywhere confirmed, after I had travelled nearly 2000 miles upon them in all directions.

About half way between the Aroostook river and the Grand Falls, we passed a small settlement of Roman Catholic Irish, whose very failings, wherever we met with them, henceforth appeared as virtues in the eyes of my fellow-traveller, in whose honour the Woodstock charivari had been got up by the Woodstock Orangemen. These families, however, were really industrious, had good crops, and appeared to be thriving. One of the settlers, called MacLachlan, had eleven children, and a farm of 200 acres, of which 60 were cleared. He had cut 10 tons of hay, and had some of the best oats and potatoes I had seen in the province. He had been in the country four years, and had cleared all with his own hands: I suppose that means with the help of his children, for all can do something. He said an emigrant who had £20 in his pocket, after paying for his land, would *be easy*. He only required a little to carry him on till his first crops were gathered. His own 200 acres, with the 60 cleared, he said, might now be worth £100. There were many excellent hard-working Scotch and Irish farmers in the neighbourhood, he added; the natives—native-born he meant—were too lazy, and liked lumbering better. Indeed, the more I saw of North America all over, the more I was satisfied that an indolent man will do better at home than on the new continent; but industry and patience are sure to be rewarded with competence and a comfortable living.

Another Irishman had been three years in the country, and a third only one year. All were happy, and had excellent crops, with new-chopped land burning for those of next year. One of these had paid £50 for his 200 acres, because a little of it had been cleared. The Government price is 3s. currency an acre, and 3d. for surveying, payable in four instalments, or 20 per cent discount for ready money; so that 1000 acres would cost £120 to Government, and £12 to the surveyor.

These Irish settlers struck me as representing industry personified. I saw many others of the same nation, afterwards, of whom I could not speak so well. The labour they undergo appears severe; but I am told, by those who have themselves gone through it, that it is not really so severe as it appears to be, and that it is by no means unpleasant. This is intelligible enough after the anxieties and seasoning of the first year are over, and the crops on the new land begin to ripen. One comfort certainly attends it, the greatest of all earthly ones, undisturbed good health. Ague and fever, as I have already said of the sea-coast of the province, are unknown; and a healthier set of children of all ages I have never seen anywhere than greet the eyes of the stranger all over this province.

The slate rocks towards this upper part of the St John become more calcareous, and beds of limestone occasionally occur, which will afford an additional means of advancement to the future agriculture of the country.

The town of Colebrook is prettily situated, on a little peninsula, formed by a sharp turn of the river St John, which here precipitates itself perpendicularly over a ledge of slate rocks from a height of 58 feet. It then proceeds through a narrow rocky gorge of hard slate for about three-quarters of a mile, in the course of which it descends 58 feet more, making its total descent 116 feet. As a picturesque object the falls are very striking, when seen from the high over-hanging rocky cliffs, and well deserve a visit. Economically, they form a great reservoir of mechanical power, which on some future day will, no doubt, be made available for useful purposes. Some years ago saw-mills were erected upon the edge of the falls on a large scale, and expensive constructions made by the late Sir John Caldwell, which brought many people about the place, and for a time quickened the growth of the town. These works, however, have been long ago abandoned; the buildings have been allowed to go to decay, and only a few rare trees were being cut up, by this huge force, when I visited the scene of Sir John's indefatigable exertions, and expensive ingenuity.

Coleridge, being the lower limit of the navigation of the Upper St John, which drains an extensive and improvable country, must hereafter become a town of considerable consequence. This will be hastened and increased if the proposed improvement in the St John, between the head of the tide-waters near Fredericton, one hundred and twenty miles below the Grand Falls, be carried into effect, and if, by means of a canal through the peninsula at Coleridge, the navigation of the upper can be connected with that of the lower part of the river. It is unfortunate that, in a new country like this, there is always more to be done than there is of money to do it with; and that, consequently, many most desirable improvements are obliged to stand over, till more favourable times arrive. Colebrook is a very old military station, which it is now thought expedient to strengthen, from its proximity to the American boundary as fixed by the Ashburton Treaty.

CHAPTER III.

Upper St John.—Colonel Coomb's farm.—Growth and consumption of buckwheat.—Aversion to the oat among settlers of French extraction. —Valley of the Madawaska.—Edmonston, or Little Falls.—Houses of the Acadian farmers.—Tea-dinners.—Ascent of the river Tobique. —Rich upper lands of this river.—Large growth of buckwheat.—Why buckwheat is unfavourable to good husbandry.—Terraces of the St John River. Autumnal tints of North America—Ferry farm at Woodstock.—Time of growth of grain crops in New Brunswick.—Sumach trees.—Apple orchards.—Scotch settlement.—Making land at Fredericton.—Rising of stones under the influence of the frost.—Turnip culture in the province.—Fire-weeds and Canada thistle. —Stanley, the settlement of the New Brunswick Land Company. —Heavy wheat in this province.—Price of farms.—Hop culture.—Running fire in the fields.—Bilbery swamp.—Farm and opinion of an Aberdonian.—Advice to intending emigrants.—Wild raspberry.—Raspberry hay.—Mare's-tail cut for hay.—Boistown.—Great fire of 1825.—Gloomy landscape.—Fires in the forest.—Nakedness of the cleared land.—An Irish settler.—Evil of farmers engaging in the timber trade.—Deserted farms, and emigration to the United States, how brought about.—Success of farmers in New Brunswick, who mind their farms only.—Price of farms on the Miramichi River.—Increasing consumption of oatmeal.—Legislative bounty for the erection of oatmeal mills.

Monday, 20th August.—At nine in the morning we started for Edmonston, or the Little Falls, at the mouth of the Madawaska, where the latter river empties itself into the St John. The distance is about forty miles. After ascending the right bank about a mile, we crossed the river by a ferry-boat, and continued our journey up the left bank, as only a few miles farther up the state of Maine comes down to the water's edge, and the river forms the international boundary. The soil and country, after we crossed the river, immediately became of inferior quality, and the settlers appeared to be both needy and indifferent cultivators.

They were chiefly French Canadians, brought here to work at the saw-mills; and who, seven years ago, on the failure of this employment, squatted on the pieces of land they now occupy. Freehold grants of land on the Upper St John were withheld by the Government, till about a year ago, when the disputed boundary question had been settled.

At a distance of twelve miles we came to Colonel Coomb's farm, the first piece of good land of any extent, upon the bank of the river, which we had yet passed. The hill-tops on each side of the road and river were generally covered with soft wood; but farther inland the land was said to be better adapted to farming purposes. It is generally upland of second quality, a sort of third-rate soil.

Colonel Coomb's farm contains 1025 acres, of which 80 only were cleared. Of these, 50 acres consist of high intervale or terrace, of a light-coloured clay loam, occasionally sandy, as is the case with nearly all the higher terraces. This intervale land he valued at £15 an acre, the cleared upland at £3, and the whole farm at £1200 to £1500. On the intervale I walked through beautiful crops of potatoes, oats, and Indian corn. The heads of the Indian corn were large, and fully formed, but had not yet escaped from their sheath. It was sown on the 28th of May, and the crop I saw would yield 50 or 60, though the average is only about 30 bushels an acre. It generally ripens here.

On the poorer soil of the upland, buckwheat is sown, and yields 35 bushels. This grain has been everywhere very extensively cultivated in New Brunswick of late years, and since the wheat has become so precarious a crop.

We saw large breadths of it, on our way up the valley, during the remainder of this day's journey; and, subsequently, in nearly all parts of the province. Colonel Coomb assured us that at least three-fourths of all the bread consumed in this district was made of buckwheat. It is used chiefly in the state of thin cakes, called pancakes. These are generally small, and, when nicely made and browned, very much resemble our English crumpets, with half their thickness. They are eaten hot, and generally with butter and molasses, or maple honey. All over Northern America these pancakes are seen at the breakfast and tea table, and are really very good. As to the nutritive quality of this grain, I find by analyses, which I have since had made, that buckwheat flour possesses about the same value, in this respect, as our best varieties of British-grown wheat.

Potatoes yield here 250 bushels an acre, and oats 30 bushels. Wheat used to yield 25 bushels. Newly cleared upland will yield 20, and old upland 10 to 15 bushels of wheat, when this crop succeeds; but for the last seven years Colonel Coomb's had not raised enough for his own family.

I found that in this valley, as I subsequently found in Lower Canada and in the north-eastern parts of this province, the oat is generally disliked as food by the natives of French extraction. This is one reason why they live so much on buckwheat cakes, and on bread made of mixed buckwheat, barley, and rye. The oats of New Brunswick are very good, and are said sometimes to weigh as much as 50 lb. a bushel. They form one of the most certain crops of the province; and hence both the cultivation and the use of the oat for food has, of late years, been greatly extending.

The oat is a kind of grain which differs much in quality and in palatableness, according to the variety raised, the climate in which it is grown, and the way in which it is manufactured into meal. For these reasons, English oats and oatmeal are generally quite different from, and inferior, both in quality and flavour, to those of Scotland; and hence one reason of the dislike which many profess against an oatmeal diet. Thus the definition of Dr Johnson, instead of being unjustly regarded as a bit of ill-natured satire, should be considered rather as the expression of a wise opinion, in which he was before his time—that, "while Scottish oats were food for man, English oats were only food for horses." As elsewhere in the province, the land on the Upper St John is generally ill-treated,—the take-all-and-give-nothing system being pursued, partly from ignorance and partly from idleness. The old Acadian French, who are settled in numbers in the upper part of this valley, are described as fine industrious men; but the Lower Canadians, who came across from the shores of the St Lawrence, are represented by the English settlers as a "miserable set." This probably arises from the fact that, as the Irish do with us, the poor Lower Canadians come into and through the country as beggars in great numbers.

There was little change in the character of the country till we were more than half-way to Edmonston. The upland was covered with soft wood, rare clearings, little rich intervale land, and that chiefly at the mouths of the small streams which come into the St John from the north. But beyond this the country improved much, both in beauty and fertility. The river channel opens up into a wide valley, with extended cultivation, scattered farm-houses, and a striking back-ground of mountains towards the north. For the last twelve miles the river had become the boundary—the one bank being British, and the other American, as it is usual to express it. This beautiful valley, with the rich lands which border the river above the mouth of the Madawaska, as far almost as that of the river St Francis, is the peculiar seat of the old Acadian French. Driven successively from one settlement to another, during our struggles on the American continent, the original French settlers in Nova Scotia had finally found refuge in this remote district. They are four or five thousand in number; and as they occupied both sides of the river, a portion of them were transferred to Maine, when the river became the boundary. They are all Roman Catholics, and have three large chapels in the district.

It was pleasant to drive along the wide flat intervale which formed the Madawaska Valley, to see the rich crops of oats, buckwheat, and potatoes; the large, often handsome, and externally clean and comfortable-looking houses of the inhabitants, with the wooded high grounds at a distance on our right, and the river on our left—on which an occasional boat, laden with stores for the lumberers, with the help of stout horses, toiled against the current towards the rarely visited head-waters of the tributary streams, where the virgin forests still stood unconscious of the axe.

Twelve miles below Edmonston, we dined at Kelly's. We found him selling mixed white and black oats to the lumberers at 2s. a bushel. He said 1s. 3d. a bushel would pay the cost of raising them. A little farther on, we crossed the mouth of the Green River, which comes in from the north, and opposite to which, on the American side, we saw the first of the large Roman Catholic chapels of the district, beautifully situated at the foot of a cliff. Within four miles of Edmonston, we passed the Madawaska chapel, a large and fine old building, with a comfortable *presbytère* close by; and soon after we met the cheerful-looking, fatherly, fat old priest driving home in his gig. About eight in the evening we reached Edmonston, or the Little Falls, which was the limit of my travels in this direction.

The river Madawaska here comes in from the north.

Its banks are settled for about twelve miles above its mouth, along the road to Canada. From Edmonston to the mouth of the Rivière de Loup on the St Lawrence, is about seventy-six miles. Along this line of road, the provincial mails are carried, and it is the usual route of travellers from the one province to the other.

Edmonston is a small village, with a comfortable inn, which will hereafter, from its position, rise into consequence. Being so near the border, it is an important military position, and is therefore nominally protected by a block-house, perched on the top of a rock. Above it, on the St John, there is some rich intervale land. I walked out to one farm upon this land, and found beautiful crops of wheat, oats, barley, buckwheat, and potatoes. The upland, immediately bounding the valley, is rocky and forbidding; but, as on the lower part of the Madawaska district, the land is said to improve on going farther inland. This inland country, however, is at present inaccessible for want of roads.

August 21*st*.—At six in the morning, we started on our return, and drove half-way to the Grand Falls to breakfast. Farms in this part of the valley, with one-half cleared, may be had for about a pound an acre. That of Mr Cyr, who gave us breakfast, consists of 350 acres, with 150 cleared, and he valued it at £300 to £400. We found a new house building here by a respectable Acadian, who has hitherto lived and farmed on the American side. On the settlement of the boundary, he became a citizen of Maine, and was sent by his neighbours to the state legislature. But he is tired of the new dominion, and is building himself this house on the British side, that he may live again under provincial laws and among his own people only.

The houses of the Acadian farmers look cleaner and more comfortable without than they often do within. I here entered one of them, and found myself in a large room, in the middle of which a stove was placed, where the baking and cooking was done, round which the family sat to warm themselves, and the pipe from which ascends through and warms the upper rooms. All the other lower rooms enter from this general family apartment. In some houses a kitchen and sitting-room are formed under this first floor and half under ground, into which the family retire in winter. Except the stone foundation, the houses consist of a wooden frame-work, with planks nailed over these, and the exterior finished off with shingles or thin boards tacked on, so as to repel the rains and drifting snows. I found a party at dinner, eating with their boiled beef the very dark bread of mixed buckwheat, barley, and rye, of which I have already spoken. As in Lower Canada, pease, as a vegetable food, have here been largely introduced and cultivated since the wheat crop became uncertain.

Above the Grand Falls we observed no hemlock trees; and it is said that they do not grow in this upper region of the St John. This fact will probably admit of a geological explanation. Again, as to the intervale land—the low intervale is generally lighter and more sandy than the low intervale of the under part of the river. This may arise partly from the lighter and finer particles of drift being carried by the flowing waters to a greater distance downwards, leaving the sand behind, and partly from the nature of the country through which the streams descend. The large feeders—the Aroostook, the Tobique, and others, which enter below the Falls—may bring down from the softer strata through which they pass the materials that render the lower intervales more heavy in their soils, and more fertile in their produce.

Pork ham was a frequent relish to our tea-dinners and tea-teas in this part of the world; but English leather would be called tender in comparison with the hams which are the pride of the valley of Madawaska. The porkers we saw frequently grunting along were to me another reminiscence of my ancient Swedish adventures. Only, in every undesirable quality, they were a little *more so* even than my Scandinavian acquaintances. My natural-historical fellow-companion pronounced them a cross between a giraffe and a crocodile—the former having given them length of leg, and the latter length of snout.

Tea is—in this province at least—almost as constant a beverage as it is in Russia. No dinner is complete without the tea; and one of the females of the household always considers it her duty to attend, during the consumption of the potatoes and the ham and other good things, to pour out the tea for the company.

In Norway, I used to amuse myself with the perpetual *lax** everywhere set before us. It was lax to breakfast, lax to dinner, and lax to supper—here, as a witty friend of mine observed, it was "*Te* veniente, *te* redeunte die."

At Grand Falls we only stopped to dine. We then returned through the better land, and the more familiar Scotch and Irish settlements, to the mouth of the Restook river, and thence to the mouth of the Tobique, where there is a comfortable inn. To-day, as during our whole excursion, the beautiful fire-weed, *Epilobium angustifolium*, springing up with its tall stem and purple flowers, wherever the forest has been burned, and sometimes over great breadths at once, as if it were a sown crop, formed an interesting and striking contrast to the blackened stems and twigs among which it grew.

22*d August.*—The mist still rested thick and heavy on the waters of the St John, flowing here at the rate of eight miles an hour, when at five minutes before five, after a hasty breakfast, my Mellicete Indian boatman, John Francis, pushed off his bark-canoe to pole and

* Dried salmon.

paddle me up the Tobique river with all speed, as far as the Red Rapids.

The Tobique and the Restook, or Aroostook, are both large feeders of the St John, descending to it, as I have already mentioned, from opposite directions. The former comes from the north-east, and is derived from four main sources, which unite into one stream about eighty miles above its confluence with the St John. The interior country through which it flows is still unsettled, with a few scattered exceptions, and inaccessible for want of roads. At the time of my visit the waters were low, and the river full of shallows and rapids, which made the ascent fatiguing, and condemned my boatman to the use of his pole for the most part, instead of his paddle. He pushed me willingly along, however, and the mist gradually cleared away as we ascended. After a couple of miles' polling, we came to the narrows of the Tobique, where the river is hemmed in by high rocks, and runs deep and swift through a chasm nearly a mile in length. There was to me a new interest in feeling myself heading in so frail a canoe these now sullen waters, and now swift and tumbling rapids, and warily avoiding the projecting rocks and rocky islets. Like the salmon underneath, the canoe seemed to pick out for itself, as if by instinct, those places in the rapids which were the easiest and safest to ascend. It was beautiful to see with what skill and strength of arm it was propelled, equally through the strong silent streams and the troubled and noisy currents.

When we emerged from the narrows, and had overcome the rapids above them, the river opened out, and presently the sun threw some of his rays slantways through the mist along the trees on the right bank of the river. Seen through a veil of unilluminated mist, the mixed pines and birch and maple thus lighted up, on the edge of the expanding sheet of water which lay between us, gave the scene, as we emerged from the gloom of the narrows, an exceeding beauty. While on the opposite bank, though still lying in the shade, the hemlock and cedar trees, with their long waving locks of hoary lichen —which selects these trees as favourite spots for its parasitical growth—contrasted strikingly with the dark-green foliage of the towering spruce and the lighter hues of the white birch.

For five miles we passed through the Indian reserve, which, as I have said, amounts in this place to about 16,000 acres. Much of it is good land, though soft wood prevails on the river banks, above the fringe of yellow birch and maple, which usually skirts the margin of the stream.

The slate rocks, as we ascended, were usually highly inclined, and covered with a thick coating of drift. The angles of inclination, however, where they became visible, after we had made some progress up the river, appeared to lessen, and occasional indications of nearly horizontal beds were seen.

At the upper end of the Indian reserve, on the right bank of the river, beneath the site of a small saw-mill, the slate rock appears rising about twenty feet above the river bed, and dipping down the stream at a high angle. But, above the section of rock, a deep bed of what appeared to be rolled slate-drift fills up the break—as if the ledge of rock had arrested it while moving; and a little above this again, the gravel bank consists of about twenty feet of this slate-drift, overlaid by about six feet of red sandy drift; and thin traces of this reddish covering are spread over the surface at considerable distances from the sandstone country to which the ascent of the Tobique conducts us. If these two layers of drift were deposited by the agency of the same current running in the same direction, they ought to be differently disposed—the red sand below, and the slate-drift above—as the newer red sandstone rocks would be carried away before the water could touch the slates which lie below them. They indicate, probably, a change in the direction of the current, by which the water before it reached this spot was made to traverse the red sandstone region, and strew its spoils over the previously distributed debris of the slates.

Above this point a few small clearances appeared, and among these one upon a little intervale on the left bank, the scene of John Bradley's farming and clearing operations. At the foot of this intervale nearly horizontal beds appeared, for the first time since I left the neighbourhood of Fredericton—beds of red sandstone, of which this was only a little apparently isolated outlier. Two miles above, at the Red Rapids, so called from the colour of the rock which forms them, the red sandstone basin begins. It consists here of red sandstones and marls, resting on the edge of the slate rocks. These red rocks extend to a distance of thirty miles up the river, being intermixed about half-way up with interstratified cliffs of gypsum. They probably belong, therefore, to the gypsiferous red sandstones, which in Nova Scotia lie immediately under the coal measures. Nearly the whole of the region, however, over which they extend, is a virgin wilderness—covered however, in many places, with varieties of hardwood timber, which are known to indicate good land. When roads shall be opened into it, I infer, from the nature of the formation, that, except where ungenial drift covers the surface deeply, this will prove one of the best farming districts in the province.

At the Falls, a large clearing existed—a good house, large barns, some land in cultivation, and the ruins of what had been an extensive and costly saw-mill establishment. Like many others in the country, this establishment had failed and gone to ruin, and the house and land were in the market. The spot was far from the world, and, for want of roads, almost inaccessible, except by the river, which at this season was barely deep enough for my bark-canoe. The Provincial Government, however, will by-and-by open roads along this river, and arrange with the Indians for the sale of their grants, when the stream of emigration is sure to direct itself up the banks of the Tobique.

I found the farm in charge of a Canadian, who had been employed as a workman in the mills. He held the 100 acres of more or less cleared land on condition of paying to the proprietors one-half of the produce of hay. I came upon him in a hollow while he was sharpening his scythe, and was attracted by what at first appeared to be a quiver suspended across his shoulder. On a nearer approach, however, it proved to be a roll of cedar bark, so rolled up as to resemble a quiver, which was lighted at one end, and attached across the shoulders in such a way as to cause the smoke from the burning bark to float about the head of the wearer. This was to keep off the flies. It is an Indian mode, I believe, in common use in the country; and on the Tobique, at certain seasons, it is said to be indispensable. The flies are most troublesome in the evening; and I had already elsewhere on the St John seen fires kindled in the open air for the benefit of the cattle, which are happy to come in the evening and hold their heads in the smoke, with the view of escaping to some extent from their tormentors. As the country becomes cleared, the flies may be expected to diminish.

The river looked very tempting above the Falls, but I had no time to ascend higher; I therefore again embarked in my canoe, and descended swiftly to the St John. My Indian took nearly five hours to pole me up, and he worked very hard; but we descended in about half the time. One soon acquires confidence in a bark-canoe, even in rough rapids; and it is certainly very interesting to observe the ease with which the sharp eye and practised hand of an Indian boatman keep it in safe waters. A little touch upon a rock, at a well-calculated time and place, snatches you from undoubted danger; and the ease with which all mishaps are avoided is apt to make the stranger fancy that it is the simpleness of the work, and not the skill of the workman, that bears him so confidently along. A single trial of his own powers, however, soon sets him right on this point.

At the mouth of the Tobique I joined my fellow-travellers, and started on my way back to Woodstock. We kept along the banks of the river all the way, and saw some fine patches of intervale land, the sites of good farms. On some of these, buckwheat, during the last few years, has been grown in very large quantity. I was told of six or seven hundred acres being raised by one individual.

This grain, I have said, is sufficiently nutritive. Those accustomed to the use of it even say that it gives more strength than any other food. In the form of cakes, the only form in which I have eaten it, it is also very palatable. But the objection to it as the staple food of a people consists in the ease with which it can be raised, the rapidity of its growth, the small quantity of seed it requires, the slovenly and unskilful husbandry which is sufficient in favourable seasons to secure average crops, and the casualties to which the crop is liable from the seasons. It grows on very poor land, from which no other grain crops in remunerative quantity can be obtained, and it is rarely favoured with the luxury of manure. Like the potato, therefore, it induces an indolent, and slovenly, and exhausting culture. And supposing the crops to fail, as the potato and the wheat have done, the poverty of the land, and the want of skill in the farmer, will render it very difficult to replace it by other crops, which demand more industry, more skill, and more attention to the collection, preservation, and application of manures, and which will refuse to grow on exhausted land.

Some of the facts above stated may, however, be considered as arguments in favour of the cultivation of this grain; and, where the soil is naturally poor, buckwheat is really a precious gift of nature, by which subsistence may be raised until, by cultivation, the land is made capable of producing more desirable crops. But it is the prelude of evil when a kind of food which requires little exertion to obtain it becomes the staple support of a people. They are sure to become indolent, and careless of further comforts. And if the food be one which, like buckwheat, will grow upon a poor soil, they are apt to allow the soil to become poor, because it will still grow this crop. Thus, they are inevitably exposed to periodical accessions of scarcity or famine, and by these visitations are certain to be reduced to permanent poverty.

At all events it is well-known that, in those parts of Europe where buckwheat is the staple food of the people, ignorance and neglect of good husbandry prevail, and great poverty is seen. In France this grain is regarded as the symbol of agricultural misery, and of the most detestable culture; and, with the chesnut, is the "triumph of improvidence and idleness." * Of the whole surface of France, less than an eightieth part ($\frac{1}{81}$) is sown with buckwheat; but in the province of Brittany one-twelfth of the whole surface is sown with this grain. "The exceeding misery of the Breton peasant was noticed by Neckar in 1784, again by Arthur Young ten years later, and, relatively to the population of the rest of France and of Great Britain, it is as conspicuous as ever. The interior of a Breton cabin in the North Breton departments is described as a parallel to that of an Irish one—buckwheat bread being the chief sustenance, instead of potatoes." † Whether this grain be the cause, or only the attendant

* *Notes Economiques sur la Statisque Agricole la France*, p. 205.
† *Proceedings of the British Association* for 1848, p. 116.

of misery, it cannot be a desirable thing to see it become the staple food of any population.*

Another consideration which may strike the traveller through countries in which buckwheat, the potato, and Indian corn form the staple food of the people, as an important objection to their use, is the constant cooking they require. Wheat, oats, rye, barley, and even pease, can be made into bread which will keep several days, or even weeks. The rye-bread in the north of Europe is in many families baked only once in two or three months. But no method, I believe, is yet known by which a palatable bread, which can be kept for days, can be made of maize, buckwheat, or potatoes. Thus, a constant cooking is required, a constant loss of time in the household, a derangement of order and neatness, and a large waste of fuel. It is chiefly female labour which is expended in this cooking and its attendant duties, but this labour in a well-regulated household is precious, and can be fully employed in other ways, in supplying the wants and contributing to the comforts of a family. This I consider an important economical and social reason why bread-producing grains should be encouraged in a country, rather than maize, buckwheat, or potatoes.

In many localities through which I passed, in this and my subsequent excursions among the Anglo-Saxon population of the rural parts of North America, I found poor log-cabins, badly-cultivated fields — dirty with weeds, and disorderly in consequence of many neglects—which light and easy labour would rectify. And while want of

* In Spain and in Sardinia the sweet acorn, the seed of the *Quercus gramuntia* of Linnæus, is eaten as a principal food of the people in certain districts. In Spain they are buried till they lose their astringent taste. In Mara Calagonis, near Cagliari, in Southern Sardinia, the miserable inhabitants, about 1100 in number, live chiefly upon this acorn, of which they make bread. They extract the bitterness from the shelled acorns by means of wood-ashes.

hands and the cost of hired labour was complained of, females in abundance might be seen, for whom the humble log-cabin could scarcely afford a reasonably constant occupation. Except among the French, and some of the Irish settlers, it is considered beneath the dignity of a female to engage in even the lighter of those healthy field-occupations which are not disdained by the wives and daughters of our European peasantry.

The terraces which the banks of the St John exhibit in so many places were very marked in the neighbourhood of Presque Isle, about half-way between the mouth of the Tobique and Woodstock. As many as five successive elevations are occasionally visible. Professor Hitchcock states that, on the Connecticut River, he has observed in some places as many as eight or nine terraces at different elevations. He states also that they only exist where traces of ancient basins in the river are visible, hemmed in by barriers, through which the river has gradually cut its way—that they are usually not parallel on the opposite sides of the river—and that they are not level, but slope downwards in the direction of the course of the river.* My leisure for observation did not permit me to ascertain how far these generalisations of Dr Hitchcock were confirmed by appearances on the river St John.

Another interesting and beautiful appearance, with which I afterwards became more familiar, first struck me, in this day's drive. This was the exceeding variety and brilliancy of the autumnal tints, which, on the branches of the young maple and poplar trees especially, began already to exhibit themselves. From a brilliant crimson, scarlet, purple, orange, or yellow, to a dull brownish red, all conceivable shades show themselves in these American forests. And what struck me as most remarkable was,

* *Proceedings of the American Association for the Advancement of Science* for 1849, p. 148.

that a single branch of a young maple would shine out with its leaves rich in these magnificent colours, while every other branch and tree around it retained its original green. And thus, over a hill-side, or along the banks of a river, patches of brilliant flowers seemed to be arranged at intervals among the verdant trees, the breadth of which varied and the hues changed from day to day. Yet often these brilliant crimson hues continue permanently, and later in the season drifted heaps of fallen leaves might be seen, retaining still the brightest tints of colour. The autumnal landscape of Northern America, coloured after nature, must appear a gross exaggeration to a European eye which has never witnessed the inimitable splendour of the reality. The bright gold of the American elm mingles in these landscapes with the red, crimson, and orange of the poplar and the maple; the unassuming yellow of the birch and beech, the browns of the basswood and the ash, and the ochrey hues of the larch, are intermixed with, or surrounded by, the dark blackish green of the prevailing pines.

23*d August.*—On leaving Woodstock this morning for Fredericton, we drove along a rich intervale, four miles in length, to the ferry, where we crossed the river and proceeded down the left bank. While waiting for the boat, I made some inquiries regarding the ferry farm, on which I saw beautiful crops of oats and Indian corn. This farm consists of 800 acres, of which from 60 to 70 are rich upper intervale land, producing 40 bushels of oats and 50 of Indian corn, and valued by the owner at £10 an acre. The rest is upland. The owner bought the whole two years ago for £700 currency. It used to be valued at £1500, but it has been long rented to an exhausting tenant, and the cultivated part has had no manure for thirty years. The selling-everything-off system was followed, and the rent, in consequence, had gradually fallen from £100 to £40 a-year, when it was sold. This exhausting system has been, and indeed still is, as I have already remarked, the almost universally followed one in North America. Ultimate poverty is the consequence of it to the farmer's family, and finally a sale of the farm itself to some one who knows how to restore it. The old occupants then trudge farther west, buy cheaply in a new country, and again inflict the consequences of evil management on some still virgin spot.

This farm is a very promising one still, to judge by the crops of Indian corn, oats, potatoes, and turnips I saw upon it. For money-returns in this quarter the farmer looks to his butter, cheese, and pork.

The oat and barley harvest is now general on the river. The usual time of growth of the different crops in the province of New Brunswick is, for

Spring wheat, . . .	3	months	and 20	days.
Barley, ...	3	„	and 6	„
Oats, ...	3	„	and 20	„
Spring rye, ...	4	„	—	„
Buckwheat, ...	3	„	and 3	„
Indian corn, ...	3	„	and 32	„

So that, as the oats are sown ten days or a fortnight earlier than the barley—generally in the last week of April or the first of May—these crops ripen about the same time in August.

On descending the left bank of the river, we observed a repetition of the same geologico-agricultural relations as we had observed in ascending. On passing from the silurian to the older slates, the surface became more stony and difficult to till and improve. On the gneiss the stones became still more frequent; and upon the granite immense granite boulders, sometimes as large as cottages, overspread the surface, and occasionally formed a perfect pavement.

The sumach, *Rhus typhina* (?) a few trees of which I had seen of great beauty in Nova Scotia, was particularly abundant along the base of the hills, and of the rocky slopes of the older slates and gneiss rocks. In a few other places in New Brunswick, and afterwards on the trap hills near Newhaven in Connecticut, I observed the same trees growing luxuriantly. Six species of *Rhus* are described by Dr Torry as indigenous to the state of New York. The greater number of these grow naturally in rocky and stony places; but whether the geological, physical, or chemical relations of these rocks and stones have any influence in determining the choice of a natural habitat is not stated by botanists. This department of vegetable physiology has hitherto been comparatively little attended to, though, in the interest of agricultural improvement, it is deserving of especial cultivation.

We travelled thirty miles down the river from Woodstock before we escaped from the stony pine-clad country. As we approached the little river Koak, we were greeted by a change in the foliage, some hardwood ridges stretched along the upland, and along the St John rich intervale appeared. We stopped at Mr Heustie's farm, on this intervale, chiefly because he is one of the most extensive apple-growers on the river.

In the orchard of this gentleman were 1550 apple-trees, for the most part young, but in full bearing. The fruit was in general small, but of a pleasant agreeable flavour. The large delicate apples of the Hudson River, of western New York, and of the Ohio River, are not to be expected, I suppose, in New Brunswick, though the summers are hot enough; yet fruit of good quality may evidently be raised, and the cultivation for home consumption carried on, with a profit.

It is probable, I think, that the great heat of the sun is in reality a chief cause of the smallness of the fruit, hastening the ripening process before the apple has had time to swell. Its scorching effect was seen upon the fallen fruit, which was dried and altered, as if by artificial heat, on the side which had been exposed to its rays. The *ten o'clock* sun has the effect also of scorching the young trees, burning a stripe all the way down the stem, and finally killing them. The preventive is to wind a straw rope round the stem, and to let all the branches grow till it has got a rough bark. It is an interesting fact that part of a stem thus protected will thicken faster than the uncovered portion, and, when the straw is detached, will be sensibly of a greater girth.

For those who are curious in such recipes, I may state that Mr Heustie kills caterpillars on his apple-trees by boring a hole half-way through the stem, filling with sulphur, and plugging it with wood. The caterpillars disappear, he says, in twenty-four hours. For lice and other small vermin, he opens a piece of the bark, introduces a few drops of turpentine, and then ties it up again. Both remedies he pronounced to be infallible.

I here first used maple sugar to my tea-dinner, enjoyed the luxury of maple syrup or honey—which with buckwheat cakes is really excellent—and ate broiled salmon with apple-sauce. In the little garden were water- and musk melons, as well as cucumbers and squashes, without any special treatment growing and ripening, as if native to the spot.

After leaving Mr Heustie's, and proceeding some miles over a country stony still with granitic boulders, and then over a patch of grey conglomerate—an outlier of the great coal measures—we turned to the left, partly to cut off a large bend of the river, and partly for the purpose of passing through a cleared and cultivated tract, known as the *Scotch Settlement.* Here there was much cleared land of second-rate quality, but I felt it to be impossible to form a satisfactory idea of its real agricultural value. The drought had been so excessive that every grass field was burned up and browned, as I had seen them in Nova Scotia. It was so melancholy to look upon the land itself, that I scarcely regretted my inability to stop and converse with the settlers, from whose mouths I could only anticipate a lamentation over the evils which the long absence of rain had brought upon them. The reader, however, must not imagine that such droughts are common in New Brunswick. Farmers who had been forty years in the country assured me, as I have already stated, that they had never known anything like the rainless spring and summer of 1849.

Leaving the Scotch Settlement, we crossed the Mactaquac River, and then a ridge of land partially settled which separates the valley of the Mactaquac from that of the Keswick River. Along the right bank of this river, the upland is of good quality in many places. It lies upon the youngest of the slate rocks which occur in this part of the province. In descending to the river, we passed through the first purely beech forest I had yet seen on my tour. It was very beautiful, entirely free from underwood, and in many places so open that it would be easy to ride through it on horseback. The soil of a beech forest is usually good, but it is shallow, resting on a hard subsoil, and therefore is not regarded as of first-rate quality.

In the bottom of the valley, and at the mouth of the Keswick River, there is much good low intervale and island land—a sandy loam, light in colour and easy to work, in favourable seasons yielding good crops, and valued at present at about £10 currency an acre. The Keswick passes through a beautiful and somewhat bold country. For the last eight or ten miles of its course, its left bank is skirted by an escarpment of grey sandstone, which here forms the north-western boundary of the coal-field of New Brunswick, and gives a character both to the soils and to the scenery. Along the course of this river grants of land were made, in 1783, to the disbanded soldiers of the New York Volunteers and Royal Guides, and their descendants have since spread, and cleared much land towards the interior.

After baiting our horses on the Keswick, darkness soon overtook us, so that the country became invisible during the last ten miles of my ride to Fredericton, where we arrived late at night after an absence of eight days.

On this excursion I had seen many spots upon which a British farmer, with a little capital, could settle comfortably, not with the prospect of becoming rich, but of obtaining all necessary comforts, and of placing upon farms of their own any number of sons. But the wise and prudent course for a new settler to pursue is to devote a few weeks to an examination of the country in person, to look at it with an agricultural and practical eye, to consult prudent persons long resident on the spot as to the advantages or disadvantages of the various farms which are to be purchased, and thus, with due caution and deliberation, and after due inquiry, to come to a determination. The emigrant and his family will then easily adapt themselves to their new circumstances; and, instead of a temporary resting-place, as so many emigrants make of the first place they settle upon, they will find at once a permanent family freehold and a happy home.

Aug. 24th.—I remained at Fredericton only for a single day, during which I visited, among others, the farm of a young improver, Mr Reid, who was spending his money in making land, I may call it, as courageously as if he had been an unencumbered Inverness or Aberdeenshire laird. The sloping upland along this part of the river is covered with fragments of sandstone; but when these are removed from the surface, a soil comparatively free from stones is found, for a depth of from one to three or four feet.

This is ascribed by some to the fact, sufficiently curious and interesting in itself, that stones and other substances in the soil have a tendency, during the frosts of winter, to rise to the surface from depths of three or four feet. The frost penetrates in many places to three feet, and sometimes deeper in severe winters; and it is observed that stones in this upper layer gradually come up to day; so that, if a field be perfectly cleared of stones one year, a new crop will be found on the land in the course of the next year, if the soil within three feet contain any. The stakes of fences are also pushed out by the same agency; and this is a source of much trouble to the farmer. During a visit to Russia some years ago, I was told the same facts by a large landed proprietor and improver, as being seen on his estates in the direction of Archangel; but I had always doubted their correctness, until the universal testimony of my New Brunswick friends put them beyond a doubt.

The effects of the potato failure are likely in this province to be ultimately as favourable to agricultural progress as we hope they may be in Ireland. It has led to a very considerable extension of the turnip culture, with its attendant advantages of stock-feeding and manure-manufacturing, and the benefits are already found so great, that should the potato come round again, this green crop will not on that account be abandoned. The turnip thrives well; and in the most northerly parts of the province I have met with crops which would not have done discredit to a Lothian farmer. The climate makes it necessary, however, to take them up in November, and to store them in cellars, where they will be readily accessible during the frosts and snows of winter.

August 25th.—Early in the morning I ferried across the St John on my way to the Miramichi River, in the north-east of the province. The road lay along the right bank of the Nashwauk, though on the upland, and at some distance from the river. The red and white fire-weeds, *Epilobium coloratum*, and *Erichtites hieracifolius*, covered as usual every open space in the woods over which the fire had run during the previous year; showing how grateful the ashes of the burned trees and underwood are to the seeds of these plants with which the soil everywhere must abound.

When land is cleared by burning for agricultural purposes, these plants cover the ground if it is neglected during the ensuing spring; but after the land has been ploughed they disappear, and the Canada thistle and the hemp-nettle come up, and become troublesome weeds. Everywhere in the provinces, and in New England, the Canadian thistle is heard of as the pest of the farm. It is our common creeping-thistle, *Cnicus arvensis*, or *Circium arvense*, which has found a most congenial climate in North America. The common spear-thistle is a troublesome weed, but, being a biennial, can be extirpated with comparatively little labour; but the Canada thistle is perennial, has deep and wide-spreading roots, is very tenacious of life, and commits its seed annually to the winds, and thus defies the labours of individual farmers. Only a general spread of cleanly husbandry will extirpate it from a district in which it has once established itself. Even legislation is disregarded by this rebellious plant. A few years ago (1847) the New Brunswick legislature passed an act, entitled "*An act to prevent the growth of thistles*," which was limited in its application to the county of Gloucester; but the refractory weed has since only spread the more, and given increased annoyance even in the county the act was intended especially to benefit.

After a couple of hours' drive, we made a detour of several miles for the purpose of visiting the village of Stanley, one of the principal settlements of the New Brunswick and Nova Scotia Land Company. We were hospitably received by Lieut. Colonel Hayne, the resident agent of the company, who accompanied me over several of the farms and clearings.

This Company, though to its shareholders it has proved a failure, has been of considerable service to the colony, and under its present management, is capable of being of much use to intending settlers from the mother country. Incorporated by act of parliament, it bought 550,000 acres from the crown in this county of York, and has opened roads in various directions, established a resident clergyman and medical practitioner at Stanley, and promoted the settlement of many respectable emigrant families in the neighbourhood. Immediately around the town of Stanley, the land is by no means of first-rate quality, but it produces 25 bushels of wheat an acre, and 200 to 300 bushels of potatoes. The wheat is thin skinned; averages 64 to 68 lb. a bushel; 68 lb. is common, and it is said sometimes to weigh as much as 70 lb. These high weights of wheat have often been given me in different parts of New Brunswick. I suppose that the very hot summers dry the grain so much as to give it superior density.

On one of the farms I visited, I found improvements proceeding as an Englishman likes to do his work—clearing, stumping, taking off stones, and trenching all at one operation. This no doubt makes the land pleasanter to look upon, and gives it a more civilised appearance than when the stumps are left for seven or eight years to rot before they are taken out. But it costs £10 currency an acre to clear it after this manner; so that, granting this method to be as cheap in the long run, it is quite beyond the means of the mass of new settlers. The owner of this farm, himself a new settler, assured me that it was a great mistake for a person with a little capital to settle in the wilderness, with the view of clearing himself a farm, when intervale land can be bought for £10 an acre. I believe there is much truth in this, unless a very favourably situated grant of good land can be obtained. The turnips (Aberdeen yellows) were on this farm very beautiful. Sown on the 19th of July, they already covered the ground.

Two farms of 100 acres each, 30 acres cleared on each, and a small house, were offered me together for £100, or £50 for each. Another farm of 205 acres, with 120 cleared, and a really nice house on it, was to be had for £750 currency.

The hop grows well here, as I afterwards saw it in the most northern parts of the province. Though there is little local demand for the produce of this plant, it might be cultivated for exportation, and would have, in the English market, at least an equal chance with the hops now imported in large quantities from the United States.

It was very striking, on one of the farms I visited, to see how rapidly fire was capable of running along a field, burning the parched grass, and endangering the crops. Advantage had been taken of the extreme drought to burn up some stumps, when, by a sudden freak of the wind I suppose, the fire took to the grass, and spread so fast towards a field of oats that it was necessary to turn out all hands to arrest it by throwing earth on the advancing line of fire; and it was finally shut out only by yoking the horses into a plough, and hastily running a furrow between the fire and the oats.

Returning from Stanley to the main road, we passed through some fine hardwood land upon the Company's grant, well adapted for farming. It was like driving through a beautiful green lane, the narrow road opened by the Company being for the most part covered with verdure, and the shade of the lofty trees affording a grateful shelter from the mid-day sun.

We came upon the Nashwauk where it ceases to be navigable, and where the ancient Indian portage across the country to the south-west Miramichi River commences. The Nashwauk, as I have already mentioned, falls into the St John opposite to Fredericton. Along its banks there is much excellent intervale land; and for upwards of twenty miles above its mouth, the cleared lands are occupied by the descendants of old soldiers of the Black Watch, (42d,) who obtained grants here in 1783, at the end of the American war. They have among them many fine farms, but their clearings have as yet extended only a short distance into the upland.

Across the portage to Boistown—about twenty miles—the country is comparatively level, and the soil is light, sandy, stony, and often poor. Its appearance was injured by the excessive drought, and the real agricultural value of the surface has been lessened by the frequent forest-burnings. At present it is almost naked of trees, and in many places forms for miles one continued fern brake.

After leaving the banks of the Nashwauk, we crossed some miles of a bilberry swamp—in other words, a bog on which bilberries grow. Half-way, we stopped to bait; and, on indifferent land, found, among other settlers, a Mr Duncan, a Scotchman. His farm consists of two hundred acres, for which he paid, when he came here, £100, ten acres being in crop. "I have plenty to eat," he said, "but I would rather pay £4 an acre for land in Aberdeenshire than be here on my own land. A man who would make his living by clearing land, in this country, must work more like a slave than a farmer."

Mr Duncan had settled himself on an unfavoured spot, and was naturally enough dissatisfied. And many such discontented and disappointed, and therefore restless and unhappy people, are to be found in all the newly-settled countries of North America, whether in the British provinces or in the United States.

Two things, indeed, cannot be too strongly impressed upon those who are about to emigrate. *First*, That those who wish to get through the world easily—who are not prepared both for privations and for very hard work—had far better stay at home. America is not a good home for idlers. *Second*, That, if the emigrant has capital, he ought to spend a little time in looking out for an eligible settlement before he fixes on a permanent home. If he have no capital to spare, let him go to service for a season, asking moderate wages till he learn where he can hope, with his small means, most happily to place himself.

In the wilderness, on burned land, besides the fireweed, the red wild raspberry, *Rubus strigosus*, springs up in vast abundance, and especially on granite and trappean soils. At Duncan's, I found it was the practice to eat this raspberry, and store it as hay. It is a kind of famine feed; but it is very frequently mixed with the hay, and the sheep are said to prefer it to common hay.

On the marsh-lands about Gagetown, on the St John, the smooth swamp horse-tail, salt-rush, or pipe-rush, *Equisetum limosum*, is largely cut for hay, as I believe is sometimes done in Great Britain. On the St John, cattle are said to fatten upon this hay, and to prefer it to the best English hay. In connection with this fact, I may mention that the field horse-tail, *Equisetum arvense*, according to Professor Torry,* is a favourite and nutritious food for horses towards the passes of the Rocky Mountains; though in Great Britain it is not only considered prejudicial to the land—or rather a sign of something to be cured in the land—but as injurious to cattle, which occasionally eat it.†

The flowed intervale lands abound also in the sensitive fern, *Onoclea sensibilis*. Upon the Keswick River, where I crossed it in returning from Woodstock, it seemed literally to cover the soil. It is cut along with the grass, and must often form a considerable proportion of the meadow hay.

At the end of the portage we descended a steep hill to

* *Botany of New York*, vol. ii. p. 481.

† Hooker.

Boistown, on the South-west Miramichi River, which runs eastward and falls into Miramichi Bay, an inlet of the Gulf of St Lawrence. This place was formerly the seat of a thriving lumber-trade, which has now almost ceased, owing to the failure of the principal adventurer by whom it was carried on. A scattering of the population has in consequence taken place, and many individual losses and social derangements have been occasioned, which it will require a considerable time to repair and adjust. Thirteen miles farther brought us to Nelson's, through virgin forests of pine growing on a poor sandy soil. The straw of the grain crops was everywhere short, and the rain had not reached the roots of the potatoes since they were planted. It was no season for judging fairly of the capabilities of the soil.

August 26th.—After breakfast, we left Nelson's for Chatham. The country continued poor, gravelly, sandy, or stony, with occasional boulders, sometimes of granite, but chiefly of the grey sandstone of the coal measures, which extend across the province from the St John at Fredericton to the Gulf of St Lawrence.

Throughout the whole of this day's journey, the effects of the dreadful fire of 1825 were visible in the blackened stems of the tall upright dead trees, which still stood undecayed, as far as the eye could see, over the gloomy hills and flats. On newly burned land the purple *Epilobium* waved its graceful leaves and purple flowers around the blackened trunks, and concealed in beauty the scorched underwood and fallen branches. But on these old burned lands the desolation was more complete, and a more sullen gloom still rested over the doomed surface. The substance of the soil is gone, it is said, where the burning has been too severe. The vegetable matter, I suppose, is consumed; and this, where no living trees are shedding their annual leaves, it must take many years to restore. Many striking facts were told us regarding this great fire, especially as to the fearful rapidity with which it hurried on, with the roar of a great sea, before the sweeping hurricane which propelled it. On our way we saw fires burning in the woods in many places, which, in this dry season, only required a little wind to spread in one blaze over the whole forest. At one spot, where the road ran along the edge of the forest, separating it from the cleared land, which lay between the road and the river, we passed six or eight men employed in watching for the fall of sparks, and extinguishing any which might come over from the burning woods, to the imminent danger of their crops.

In a country like this, one learns to look upon trees in a new light. Not only are they an obstacle to cultivation, which must therefore be cut down and burnt; but, so long as natural woods are near, it is dangerous to leave any about the dwelling-house for shelter or ornament. During this summer's tour, I was shown places where the spreading of fire from the forest to a few ornamental trees had caused the destruction of the whole farm buildings, to the almost total ruin of the proprietor. Thus a reason appears for the nakedness which an Englishman almost feels when in the midst of a large clearing. An unsheltered house appears, while the stumps of magnificent trees all around show how well it might have been protected from wind and sun.

Except upon the immediate banks of the river, there are few settlements along this road; and, in general, the upland is very poor until we descend to within twenty or thirty miles of the mouth of the Miramichi. About a dozen miles from Boistown, I had a conversation with a small farmer, Irish by birth, but resident from his infancy in this country. He had been in his farm only three years. By hiring himself as a working lumberer, he had saved £80, and with this he bought his present farm. It contains two hundred acres, and had ten acres cleared upon it, and a small log-house, but no barn. He has built a barn and added to his clearing, and if seasons come round, he should do well.

We passed houses and clearings, however, which were altogether deserted. This was partly owing to the failures in the crops, which have ruined so many of all classes in Ireland as well as here; partly to the failure of the lumber-trade, and to the debts and mortgages in which the small farmers, by engaging in this trade, had gradually become involved.

A stranger does not readily comprehend how a depression in the lumber-trade should seriously affect the interests of the rural population in any other way than in lessening the demand for produce, and in lowering prices. And it was not till I had been longer in the country, and conversed with many persons on the subject, that I was enabled clearly to separate, in my own mind, the evils which this trade had brought upon the rural population from those which were necessarily attendant upon the calling of a farmer.

In lumbering, a man goes into the woods in winter, cuts down trees, and hauls them to a brook, down which, when the spring freshets come, he can float them to the main river, and then to the saw-mills of the merchant to whom he sells them. If a man does this upon his own farm, or at no great distance from it, and by the aid of his own family only, all he gets for his wood is pure gain—if, in the mean time, he has been living on the produce of his own farm.

But if he goes to a distance from his own farm, and has been obliged to hire labourers, or has done so with the view of enlarging his operations, he must apply to the merchant for an advance of stores adequate to the winter's consumption. The cost of these stores, and the wages of his men, are deducted from the value of the wood he has obtained; and if the price of wood be not very low, he may still have a handsome surplus.

Such circumstances lure him on till an unfavourable winter comes, and he is not successful in cutting as good lumber, or in as large a quantity as usual, or in hauling it to the floating place; or a very late spring, or very shallow water, prevents him from getting it to market. Then his debt to the merchant for stores, and for money to pay his men, must stand over to another year; and his farm is mortgaged as security for the payment.

Meanwhile this farm has been more or less neglected, and has been every year growing less produce. His wood must be floated in spring, when his crops ought to be put into the ground. He has been absent in winter, when new land might have been cleared. His mind is occupied with other cares: he does not settle to his agricultural pursuits, and they are therefore badly conducted, even when he is at home to superintend them. And, lastly, while living in the woods, both employer and employed live on the most expensive food. They scorn anything but the fattest pork from the United States, and the finest Genessee flour. The more homely food, therefore, which their own farms produce, becomes distasteful to them; and thus expensive and sometimes immoral habits are introduced into their families, which cause more frequent demands upon the merchant, and a consequent yearly increase of the unpaid bills.

In such a state of things, the foreclosing of mortgages, the sale of farms, and the emigration of ruined families, must necessarily be of occasional occurrence. But if the price of lumber fall very much at any period, they must become more frequent; or, if a merchant who holds many of these mortgages himself fails, a common ruin will involve all. Both of these evils have at once befallen the lumbering farmers on the Miramichi, and much distress has been the result. To this cause was owing the abandonment of farms by persons who, leaving both debts and mortgages behind, and taking with them any capital they could secure, had moved west to lumber on the Aroostook, or to begin life anew in the far off Wisconsin.

I have been thus detailed in my observations upon this subject, because I felt myself inclined to be unjust in my judgment as to the agricultural capabilities of a district from which so many were emigrating, and in which land was so little esteemed that its owners appeared to be abandoning it, with all their improvements, merely because it refused to support their families. A knowledge of all the circumstances, however, satisfactorily showed that not the land, but the haste of its owners to become rich, and their discontent with the slow but certain gains of agriculture, were the causes of the distress from which so many of the farmers were suffering.

With the view of obtaining a more general body of testimony in regard to the agricultural condition of New Brunswick, I was enabled, through the kind co-operation of the Provincial Government, to circulate a set of queries among the owners of the land in every county of the province. One of these queries referred to the profits of pure farming; and the numerous answers I received were unanimous in declaring—"*That, in every part of the province, those who for a series of years had confined their attention to farming alone, had all, without exception, done well.*" That prosperity, therefore, may attend the new settler, or may return to the older farmers of the province, it is necessary only that they confine their attention solely to the business of their farms.

On the Miramichi, at present, owing to failures and the foreclosing of mortgages, land is cheaper than it has been for many years. At Bergoris, twenty miles from

Boistown, where we stopped to bait, the landlord told me of a farm on the river, containing 1500 acres, having 60 cleared—and of these 20 were intervale land, producing 30 to 35 tons of hay per acre—which could be obtained for £150 to £200. Five years ago this farm would have brought £400 or £500.

A few miles farther on, after passing the mouth of the Renous river, which comes in from the left, the land became of better quality. Though we were still upon sandstones of the coal measures, and the surface stones were chiefly sandstone boulders, sometimes mixed with frequent masses of granite, yet the soil was more tenacious and clayey; and good crops of wheat and oats were ripening upon many of the wayside farms we passed.

On the Miramichi we looked in vain for the frequent fields of buckwheat, which we had seen upon the St John. The oat here takes its place, and is gradually assuming an important place as an article of ordinary diet among the people. Until lately, the humblest people refused to eat anything but the finest flour. They even thought they could not live upon anything else. But the failure of home wheat, and the want of money to buy that imported from Canada or the United States, has had the salutary effect of compelling the people to try the virtues of their own excellent oats; and it is to be hoped they will every year become more and more attached to this most nutritive grain. The Provincial Legislature have most judiciously aided this alteration by offering bounties for the erection of oat-meal mills throughout the province, the want of which had hitherto been almost a complete bar to the use of the oat as human food—especially in the newly settled districts, where the need was most urgent, and the want most felt. In 1847 the sum of £500 was paid out of the Provincial Treasury for this purpose; and as such mills are not costly, the wants of many districts have been already fully supplied.

About sunset we reached the ferry across the north-west, where it joins the main or south-west Miramichi river, and travelled the remaining ten miles to Douglas chiefly in the dark. The land is generally of better quality along this lower part of the river, is more extensively cleared, and more skilfully cultivated. Newcastle, a considerable village four miles below the junction of the north-west river, and Douglas, a town six miles farther down, are supported in part by their traffic with the country farmers, but chiefly by the lumber-trade, of which the mouth of the Miramichi has long been an important centre.

Soon after leaving Newcastle, we met with an accident by which the pole of our carriage was broken—a circumstance of the more importance as we had still some hills to descend before we could reach Douglas.

But my travelling companion, Mr Brown, was equal to any emergency. A spare rope, and a couple of stakes from the fence, in his hands soon placed us again in marching order; so that, with a little care, and by walking on foot down the dark slopes, we reached Douglastown in safety before midnight.

CHAPTER IV.

Douglastown.—Great heat.—Value of farms.—Mode of reclaiming forest land.—Expense of clearing soon repaid.—Plague of grasshoppers, in New Brunswick and New England. — Legislative grants for the promotion of agriculture. — Average produce, prices and wages in Northumberland county. — Town of Chatham. — Golden rod, a troublesome weed.—North American oaks.—European weeds on the cleared lands. — History of an Annandale settler. — Bush-bean.—Provincial encouragement to elementary and grammar schools. —Bay-du- Vin schoolmaster.—Richibucto.—Buctouche river.—Sweet fern soils — Patience and contentment of the French settlers, and restlessness of the Anglo-Saxons.—Shediac, famed for its oysters.—The Bend-Bore of the river Petitcodiac.—Height of high water above that of the Bay.— Country between the Bend and the city of St John.—Case of Mr Nixon, and his opinion of New Brunswick as a poor man's country.— Use of river mud as an improver of the soil. — Greater industry among new settlers than among the native-born.—Blighting of buckwheat.—Burned Bridge.—Beauty of Sussex Vale.—Mr Evanson's home farm, its value and produce.—Mr Aiton's farm.—Rent and course of cropping. — Hampton, and its conglomerate soils. — Fine-looking yeomen of New Brunswick.—Price of farms around Hampton.—A discontented Irishman.—Dyked marshes of St John and the Atlantic border. — Farms around St John, their quality and value. — Rate of wages for agricultural labour in the several counties of the province.

August 27.—Yesterday and to-day have been excessively hot. We found it so as we travelled down the river in our open carriage, but we had no means of ascertaining the temperature. At Douglastown, I am informed that the thermometer has frequently stood during the past week as high as 95° Fahr. in the shade.

On this, and a subsequent visit to the Miramichi, I was much indebted to the hospitality of Mr Rankine, one of the oldest resident merchants, and the representative of a wealthy firm long connected with the North American colonial trade. I visited with him to-day the farm of Mr John Porter, on which I found good land, well cultivated, with fair crops of wheat and oats, and a field of excellent turnips, (Aberdeen yellows.) The wheat averages 20 bushels per acre, of 60 to 65 lb., and the oats 40 bushels of 37 to 40 lb. On the upland, where the soil is heavier, the oats weigh as high as 48 lb.

This farm is mostly flat land—an extension of the high intervale on which the town stands. It consists of 80 acres, of which 60 are cleared, and is worth £400, but would at present sell for £300. He assured me that, though he has a large family, he could make a living off this farm.

Above this, the same gentleman possesses another farm on the upland. It is stronger land, and produces better oats; but it is more difficult to work, and is later in spring. It consists of 150 acres, of which 50 are cleared, yields 15 tons of hay, lets for a money-rent of £33, and is valued at £400, all currency.* Ten years ago, this farm was let for £50. The tenants have never done anything else but farm, and they have been enabled to support their families and pay their rents — though, as I have already remarked, the renting of farms is not a popular or much practised mode in this country. It is an excellent plan, however, for a new beginner, who wishes to know something of the country before he fixes upon a spot for his permanent residence. Much of the moving, and of the want of local attachment which is seen in North America, is probably to be ascribed to the hasty settlement which circumstances compel so many emigrants to make on their arrival in America.

The course of cropping adopted by a skilful man like

* £20 sterling make £25 provincial currency.

Mr Porter, on clearing new land from the forest, will give the reader an idea of the general character of the treatment to which less prudent men subject their land. He cuts down the wood and burns it, then takes a crop of potatoes, followed by one of wheat with grass seeds. Nine successive crops of hay follow in as many years; after which the stumps are taken up, the land is ploughed, a crop of wheat is taken; it is then manured for the first time, or limed, and laid down again for a similar succession of crops of hay. This treatment is hard enough; but the unskilful man, after burning and spreading the ashes, takes two or three or more crops of grain, leaves it to sow itself with grass, then cuts hay as long as it bears a crop which is worth the cutting—after all which he either stumps and ploughs it, or leaves it to run again into the wilderness state.

In clearing land in this district, it is calculated that the first three crops, which are merely harrowed in, will pay all the expense of cutting the timber, burning, and cultivating. If the settler then abandon it, he is no loser: everything he cuts off it afterwards is gain, or any sum for which he can sell his cleared land. This is a great inducement to the exhausting system, which clears annually new land for grain, cuts for hay all which the old cropped land will yield, till it is again overrun with a young growth of wood, and neither saves, collects, nor values manure.

This system is barbarous, reprehensible, and wasteful to the country—and yet it is probably the method which yields a ready sustenance to the settler's family at the smallest expense of mental and bodily labour. Our condemnation of the pioneers of civilisation in a new country ought not, therefore, to be too severe or indiscriminate. With all our skill, we English farmers and teachers of agricultural science should, in the same circumstances, probably do just the same, so long as land was plenty, labour scarce and dear, markets few and distant, and prices of produce low. As population increases, a higher class will come in, will purchase the exhausted farms, and for their skill and manure will obtain from the soil new returns, as large, and perhaps as profitable as those which rewarded the men who first penetrated the bush. Or if such men do not come in, and the land still continues in the hands of the original clearers, or their sons, the good of the country will demand that steps should be taken to instruct and enlighten them in regard to the principles of agriculture, and by degrees to wean them from an agricultural routine which is no longer either the most profitable to the individual, or adapted to the altered circumstances of the country.

In walking over Mr Porter's farm, my attention was drawn to the vast number of grasshoppers which were jumping about, not only in his grass, but in his turnip fields. I had observed them previously in considerable numbers at various places on the St John River, but here the land seemed almost alive with them. They appear during the hot weather of midsummer and autumn, and attack the turnip crops as well as the grass, sometimes entirely stripping them of their leaves. If the young turnips are not sufficiently forward by the middle or end of July, when the grasshoppers begin to swarm, they are sometimes entirely destroyed. This is a pest of which our British turnip-growers, so far as I am aware, have no cause to complain.

In New England, five or six different grasshoppers, besides as many species of locust, appear in their warm summers. In Massachusetts, the grass in the meadows and moist fields is filled with myriads of small grasshoppers, of a light green colour, which do much injury to the grass. But, in New England, grasshoppers are not generally distinguished from the small varieties of locusts which are common in that country. One of these, the small red-legged locust, about an inch in length, infests the salt marshes in such numbers as almost entirely to consume the grass; and when the scanty crop of hay is gathered, it is so tainted with the putrescent bodies of the dead locusts contained in it that it is rejected by cattle and horses.* It is some small return for their ravages that the bodies of these creatures manure the fields they have infested, and that poultry thrive upon them. Young turkeys, in the summer, live almost entirely upon these grasshoppers in parts of Massachusetts, and become fat.

The Northumberland Agricultural Society, which has its headquarters at Douglas, has hitherto been the most influential in the province, and has received the largest share of the legislative grant for the encouragement of agriculture. A method of promoting improvement among the rural population, which is common to the provincial and to the New England state legislatures, is to give from the public funds to every society a sum of money, bearing a fixed proportion to the amount raised among its own members. In New Brunswick, for every pound subscribed in a district for the promotion of agriculture, the Legislature formerly gave £2, and now give as much as £3, from the Provincial Treasury, thus stimulating at once and rewarding the local subscribers. For this purpose, £6150 were voted by the New Brunswick Legislature in 1848.

In this district I found some of the best farming and best farmers in the province, and some of the warmest friends of agricultural improvement. As there are at present many farms to be disposed of upon the Miramichi River, for which persons who know something of agriculture are eagerly desired from the Old Country, I shall insert the average produce, price, and weight per bushel, of the usually cultivated crops in the county of

* HARRIS's *Insects of Massachusetts Injurious to Agriculture,* p. 136.

Northumberland, which embraces most of the good land on the Miramichi and its tributary waters.

	Average produce per acre.	Average price per bushel.	Average weight per bushel.
Wheat, .	17	7s. 6d.	62 pounds.
Barley, .	32	5s. 3d.	54 ...
Oats, .	32	2s. 1d.	38 ...
Buckwheat,	40	5s.	45 ...
Maize, .	50	4s. 6d.	58 ...
Potatoes, .	200	1s. 10d.	
Turnips, .	350	1s. 4d.	
Hay, .	2 tons.		

The average price of cheese is 5½d. a pound, and of butter 9d. The average wages paid for agricultural labour are, in summer, from 50s. to 60s. a month, or, by the whole year, £27, 10s. currency, in addition to board, washing, and lodging. Day-labourers receive from one-half to three-quarters of a dollar, with provisions.

After a hasty survey of the neighbourhood of Douglas, we drove down to the river, and crossed to the town of Chatham, on the right bank. Near the ferry we found a large encampment of Indians, who are about two hundred strong, on the Miramichi River, and own reserves of about 22,000 acres, some of which consist of excellent land. It was amusing to see the little papooses, only a few weeks old, swaddled up tight, tied fast to a bit of board, and set on end against the outside of the wigwams, apparently unheeded by anybody. No movement was made by any of the females, nor a sound uttered by the infant, when I took up one of them and affected to carry it off with me to the boat.

The town of Chatham is about equal in size to Douglas, and, like it, is dependent partly upon the lumber-trade and partly upon the agricultural traffic. On this occasion I merely drove through it with the view of reaching Richibucto, a distance of forty miles south, before nightfall.

Half-an-hour brought us to the Napan River, a stream which widens as it descends, and falls into Miramichi Bay. On this river there is much good strong land, a stiff clay, the first I had seen in the settlement, for the improvement of which I was satisfied, notwithstanding the drought—which even here had reduced the hay-crop to one-third of its usual amount—that the system of thorough drainage might, even in this climate, be unhesitatingly recommended. This clay is specially infected with two species of golden rod, (*Solidago canadensis* and *S. altissima*,) which are troublesome weeds, and the former especially difficult to extirpate.

Neither of these species of golden rod is known as a weed in Europe. The only European species is the *Solidago virgo aurea*, which is also a native of America. It is not known as yet how many species of golden rod are to be found in New Brunswick; but in the state of New York no less than twenty-two species are known. It is very interesting to the botanist and physiologist to observe such differences in the flora of countries so closely allied as Great Britain and Northern America now are; but, as practical indications of the qualities of soils, this new flora is a source of difficulty to the visitor or settler from the Old Country, who is accustomed from early observation to connect in his mind the qualities of a soil with the weeds which grow upon it. It is a matter of regret that botanical collectors do not describe more particularly both the kind of soil on which plants usually occur—which, when troublesome weeds, they infest—and the geological formations on which they are most frequently found. A practical value would thus be given to botanical descriptions, which hitherto they have seldom possessed.

To the English traveller, who is less interested about the indications of the humbler vegetable tribes, the numerous new species of familiar kinds of trees he meets with in Northern America are more striking. Thus in

Great Britain, and in central Europe generally, there are seen only two kinds of oak, the common British and the sessile-fruited oak, *Quercus robur* and *Q. sessiliflora.* In Nova Scotia and New Brunswick he does not find these oak trees, but in their stead two others, the red and grey oaks, *Q. rubra* and *Q. borealis.* If he goes south, the number of species increases. In Massachusetts he already meets with eleven, and in New York fifteen species of oak, among which the northern or grey oak is not included. In the whole United States, no less than forty species of oak are already known.

But probably the most generally interesting fact in regard to American plants is the influence which the introduction of European races and manners, and the frequency of intercourse with European countries, is said to have had upon the prevailing weeds, especially of the Atlantic coasts and river borders of North America. The common plantain, *Plantago major*, was called by the Indians the White-man's-foot. The Canadian, or creeping thistle, *Cnicus arvensis*, or *Cirsium arvense*, the pest of North American farmers, and therefore often called the cursed thistle, is an importation from Europe. Not only have most of the cultivated plants and grasses been brought from Europe, but, according to Agassiz, all the plants growing by the road-sides are exotics.

> " And the wheat came up, and the bearded rye,
> Beneath the breath of an unknown sky."

" Everywhere in the track of the white man we find European plants; the native weeds have disappeared before him like the Indian. Even along the railroads we find few indigenous species. On the road between Boston and Salem, although the ground is uncultivated, all the plants along the track and in the ditches are foreign." *

How curious are the reflections which such facts suggest! Would it be irrational in an Indian to suppose

* *Lake Superior, its Physical Character*, &c. p. 10. Boston, 1850.

that these European weeds in the ditches of the Salem railroad had actually followed the footsteps of the Irish emigrants who dug them? May not the seeds of them have been actually shaken from the shoes of the newly-arrived immigrants?

On the Napan River there are many settlers from Dumfriesshire, chiefly from Annandale. This has arisen from the circumstance that a traffic in timber has long existed between the Miramichi River and the town of Annan, the ships engaged in which afforded an easy passage to intending emigrants. Their sons are noted as the best ploughmen, and themselves as among the best farmers in the province.

With one of these settlers, John M'Lean, I had an interesting conversation; and as his history may interest some of my readers also, as an example of the way in which steady industry overcomes difficulties, and secures comparative prosperity in a new country, I shall state the leading facts I gathered from him. He came over in the year 1822. He has 250 acres in his farm, of which 150 are cleared; but he has not force to keep all this land in crop. He works it with the aid of three of his sons, two daughters, and three horses—keeps eleven cows, eight or nine young cattle, and a few sheep. He bought his land in the wild state, cleared it all himself without hired labour, and has *raised* eleven children. He has four sons settled on farms, one of whom paid £150 for his farm; two of them worked as carpenters till they had saved money to buy their farms. Neither he nor any of his children ever lumbered, nor should any of them if he could help it. Not one in twenty makes anything by lumbering; and by sticking to their farms, men in the long run always make a better living, and are more independent, than by anything else. Many others who came out with him, and since he came, have stuck to their farms, and have done as well as himself. Though the crops have failed so many years, few in this settlement are in debt. Oatmeal porridge and milk twice a-day, and oatmeal cakes, are the prevailing diet. Odds and ends, as he called sugar, tea, &c., are obtained by the sale of butter and cheese.

Since the failure of the potato, the bush-bean—a prolific French or kidney bean, of which many varieties are cultivated in the United States—has been much grown in this district. It comes a fortnight earlier than the potato, is very prolific, and, when green, is an excellent substitute for the potato. The dry bean is usually baked with pork. This vegetable would probably succeed well in our climate, and as a substitute for the potato, if only in part, is well deserving of a trial among us.

Mr M'Lean thinks a man would do well in Northumberland, who could come over with £50 in his pocket, and better with £100. But he ought not to have too much, if he is to labour contentedly, and to prosper. He had himself only £5 when he settled, besides three carts and a year's provisions.

If these statements of Mr M'Lean are got by heart by the intending emigrant to the wilderness parts of North America, he will require little other guidance to comfort, prosperity, and contentment.

Three miles farther, over a flat coal sandstone country, brought us to the Black River, which also empties itself into the Miramichi Bay, has good heavy land along its banks, and a prosperous agricultural settlement. The next twelve miles to the Bay-du-Vin River, is over a poor sandy country, with occasional patches of cold clay and of peat bog, resting on the flat impervious sandstones.

The Bay-du-Vin Settlement consists of about a dozen Irish families, who have a school and Catholic chapel. The schoolmaster said he had forty boys at school, but the parents were poor, and paid little; so that his chief support was the provincial grant of £20 which he annually received. In regard to their elementary schools, the Provincial Legislature has lately adopted important means of improvement in the establishment of training schools at St John and Fredericton. Every settlement in the province has its school. The settlers build a school-house, and select a master. This master must then—if not a pupil of the training school, or if he has not been previously examined—undergo an examination by the master of the training school, and, according to his proficiency, he receives a certificate, which entitles him to an annual stipend of £10, £20, or £30. As yet, the training schools have not been able to supply instructed masters for all the schools; houses are not in most places built for the masters, and the schools are usually shut up in the summer months. This is very much the case also with the schools in the newer states of the Union, and as the population becomes more dense, these evils will disappear. In the principal town of each county a grammar-school is established, to which larger grants are given. In 1848, the grants to parish and Madras schools amounted to £13,882.

The schoolmaster teaches the religious catechism which the parents of his pupils wish their children to learn. Thus the same master sometimes teaches in the same school the Church of England Catechism, the Assembly's Catechism, and that of the Romish Church. The schoolmaster at Bay du Vin was surprised that I should think there was anything remarkable in his being required to teach all the three, though he said he had once before heard some one make remarks regarding it. He was himself a Roman Catholic; but it was enough for him that he had been *ordered* to do it.

We drove rapidly over the remaining twenty-two miles to Richibucto. This tract of country presented the common level surface, and the sandy, often thin, poor, and stony soils which distinguish so much of the coal measures of New Brunswick. We crossed on our way the Kouchibouguac, the Kouchibouguasis, and the Aldouane rivers, flowing from the west; and for some distance on each side of these rivers, good land and fine settlements were almost universally to be seen. On the flat country, the clearings are few and thinly scattered. Hardwood ridges of land now and then rose above the flat country, and on these were more valuable farms and settlements. But, as I came over this ground again in the ensuing October, I omit any further observations in this place.

28*th.*—I left Richibucto at eight in the morning. It was at first misty, but by degrees became excessively hot. After a few miles of better land, the soil and country became very similar to that we passed over yesterday—flat, poor, pine-clad, sandy soils, except where rivulets and armlets of the sea occurred. Here better land, and a few settlers, as usual, occurred. Fifteen miles brought us to the Big Buctouche River, near its mouth, where it expands into an arm of the sea, and falls into Northumberland Strait, opposite to Prince Edward's Island. Tide-water here extends six or eight miles above the bridge and harbour; and along the inland bay there is much cleared land, and some good farms.

On the sandy soils of the county of Kent, north and south of Richibucto, I saw, for the first time in New Brunswick, the sweet fern (*Comptonia asplenifolia,*) which had previously arrested my attention in the great sandy plain of Aylesford, in the Annapolis Valley between Windsor and Bridgetown. I afterwards saw it on many other of the more barren sandy portions of the north-eastern part of this province. It is a saying in some of our own rural districts, that, where the common fern grows naturally, something more valuable may be made to grow by art—and so it was remarked to me of the sweet fern in Nova Scotia; but except for rye or buckwheat, or the horse-chesnut, or some similar sand-loving plants, the soils on which the *Comptonia* abounded have generally appeared to me very little adapted to the economical production of vegetable forms likely to minister to the sustenance of man.

Eleven miles farther brought us to the Cocagne River, in the neighbourhood of which there are extensive clearings, and much improved land, but generally light and sandy. A shade of red had begun some time ago to appear in the soil, as if a portion of red drift from the old red sandstones of Prince Edward's Island had been transported, and mixed up with the natural debris of the coal measures of the country.

We saw a few patches of Indian corn on our route to-day, generally poor crops, unlike what we had seen upon the St John, and on the farms of the old French settlers. Along this coast the French are numerous. They have suffered much from the failures of the crops during the last two or three years, but few have left the country. They are a much more patient and contented race than the Anglo-Saxon settlers, who are always painting in brighter colours the beauty and fertility of places they have never seen, and shading more deeply the evils that surround them. Six miles beyond Cocagne I stopped at a farm of 500 acres, bought by its owner thirty years ago for £14, and now valued, with the stock upon it, to the poor-rate at £1000. Yet this man, who had so prospered, wished to sell his farm, that he might start for Indiana. He had no good reason to give for going but the same failure of crops which the humbler habitants so patiently bear. It is to be hoped that, both for Europe and America, these visitations are now for a period overpast.

From Cocagne is nine miles to Shediac, a village of some twenty houses, with a little-frequented harbour, which, among the New Brunswick gourmands, is famed for its oysters. The oysters of this coast, and of the shores of the Gulf of St Lawrence generally, are very different in size and appearance from our comparatively small English oysters. They are of a species known as the Canadian oyster (*Ostrea canadensis*,) are very large, and inhabit a shell which is long, narrow, massive, somewhat curved, and often attains a length of eight inches. The heavy shells are frequently burned into lime, and are occasionally seen in large heaps on the road-sides, collected for this purpose. I don't know to what extent it actually takes place, but I was told that the New Brunswick pigs have learned to open and relish the oysters, and that they frequent the sea-shore and contrive to feed upon them!

We had arrived at Shediac in good time for dinner; but as I wished to accomplish the remaining thirteen miles of my day's journey while it was still light, I left my provincial friends to enjoy their oyster feast, mounted a single horse car, with a young habitant for my guide, and drove on to the Bend, as it is called, on the river Petitcodiac. There were many fires to be seen raging in the woods, and the evening came on early in consequence of the smoke with which the air was everywhere filled.

The Bend is a village which derives its name from being situated at the point where the river Petitcodiac, which had been flowing north-east, bends suddenly almost at a right angle, and flows south-east towards Shepody Bay, one of the head forks of the Bay of Fundy. The tide flows up from this bay a distance of 20 miles, and rises at common tides 22 feet 8 inches, and at the highest tides 28 feet 8 inches. It rushes up with a bore which at spring-tides is 5 or six feet high, and to boatmen and shipmasters who are unacquainted with the river is sometimes a source of danger. Dr Gesner states that it is three hours' flood at the mouth of the river before the tide reaches the Bend, that it flows in and ebbs off in six hours, and, though the rise of the tide in feet is not so great as at places near its mouth, yet that the level of high water at the Bend is in reality several feet higher than that of high water at the head of Shepody Bay.*

Thus, at Dorchester Island, in the open bay, and at the Bend, the height of high water above the level of low water at the former place is as follows:—

	At Dorchester Island.	At the Bend.	Difference.
Common tides, . .	36 feet.	45 feet 4 in.	$9\frac{1}{3}$ feet.
Highest tides, . . .	48 feet.	57 feet 4 in.	

The level of the river at high water is therefore 9 feet higher than it is at the mouth of the river. Above the level of high water on the Atlantic coast, it is probably several feet higher still. Could the outfalls of the river be improved, therefore, as has been the case in our north of England river Wear, a much more perfect drainage of the river-side intervale land might be effected, and the level of high water so much reduced as to render securely dry many tracts now liable to periodical overflow.

It is to the former prevalence of such phenomena as this, and not to a real elevation of the land, that some of the discoveries of animal remains embedded in alluvial soils, now considerably above the existing level of adjoining waters, is to be ascribed.

At high water, the river at the Bend is broad, deep, and beautiful. Vessels of 100 tons can then come up to the town, which is thus enabled to carry on a direct trade with Boston in the United States. In the summer months a steamer also plies between the town and the

* Dr Gesner's *New Brunswick*, p. 90.

city of St John. At low water the river narrows exceedingly, and deep, muddy, and sandy flats appear. Along the immediate banks of the stream there is much fine land, and the mud, which is copiously deposited between high and low water marks, is very rich, and is extensively employed for manuring purposes.

29th August.—We drove twelve miles up the Petitcodiac River this morning to breakfast. From the Bend to St John, a distance of ninety-four miles, the country is more or less settled all the way. Much of the land is of inferior quality, light, sandy, and gravelly; but there is much good land also, and the country generally along the road is more undulating, abounds more in the picturesque, and has more the air of a civilised old settled region than almost any other tract of equal extent in the province. This superiority arises, in great measure, from a change in the prevailing geological formation. Before reaching the Bend, we pass from the grey coal measure sandstones on to red marls, red sandstones, and red conglomerates, with subordinate beds of limestone and gypsum, which extend to within a few miles of St John. The round hills of our English Monmouthshire appear in this region in Mount Pisgah, Piccadilly Mountain, and other striking elevations which are still unnamed. Soils like some of those in South Wales, "which eat up all the manure, and drink up all the water," are formed in many spots from the drift of these red rocks, while in not a few places red soils like those of the Lothians are produced, over which extend rich and fertile farms. Over all this red district the land is absorbent of moisture, easy to till, and early ready for the seed in spring, or for the sickle in autumn. The subordinate beds of lime and gypsum also contribute to improve the soils which rest upon this formation; while the salt with which it is impregnated, as shown by its salt springs, is probably not without its influence on the general vegetation. Along this line of country, wheat used to be grown in large quantity, for the supply of other parts of the province; but the ravages of the wheat-fly, for the last five years, have made it necessary to substitute oats and buckwheat in its place, even for local consumption.

Our landlord, Mr Nixon, with whom we breakfasted, had a comfortable house, a nice family, and certainly the very cleanest and tidiest kitchen I have seen in New Brunswick. He came from home, and settled here in the wilderness, eighteen years ago. For his farm of 275 acres, he paid £50. A hundred acres of it are now cleared, twenty-five of them being still in stump. Over and above the price they paid for the land, he and his two brothers had only £60 to begin with; but at the end of ten years the farm and stock were valued at £1000, and he bought his brothers out. The land and buildings are now, in the depressed state of things, worth about £800. He considers New Brunswick *a good poor man's country*—an expression which briefly includes all the main recommendations of North America generally to the inhabitants of Europe. "Those who are comfortable at home," as another settler said to me, "had better stay there."

The land on this farm is somewhat light and sandy; but it is greatly enriched by a covering of the river mud. Of this, one hundred loads are applied to the acre—being dug up dry during the frosts of winter, when other work is scarce, and spread over the surface either of the tillage or grass land—and its good effects last for twenty years. From his upland, by this treatment, he cuts two and a-half tons of hay an acre, obtains 300 or 400 bushels of potatoes, and as much as 1200 bushels of turnips.

It is a subject of almost universal remark throughout the province that, as a general rule, British-born settlers succeed better than the provincial-born or natives. They are steadier, more persevering, and more industrious. And I could not help remarking myself, that, in New

Brunswick as a whole, the regularly settled inhabitants did not appear to work so hard as the same classes do at home. From that fact, however, I did not feel myself justified, as some did, in concluding that the native-born are naturally or absolutely indolent; my conclusion was rather that a living was easier got in the province than in the home islands, and that, therefore, they did not require to work so hard to obtain it as we do at home.

At the same time, the absence of the strong spur of necessity does tempt poor human nature to indolence in America as it does in Europe. It is remarked among the free Blacks in the middle states of the Union, that those who have bought their own freedom are far more energetic and industrious men than the coloured people who have inherited their freedom. In Boston, it is observed that the sons of rich men rarely succeed as merchants; and no doubt there must be some truth in the statement, that the sons and grandsons of British settlers do not display the same energy as their emigrant fathers.

Whether such differences are greater or more observable in America than they are in Europe is at least open to question, though it is a point of much interest in connection with the impossibility, alleged by some, of the Anglo-Saxon race ever becoming permanently fixed and acclimatised in the New World.

The excessive heat of the season was here injuring the buckwheat, blighting what was not set, and ripening too fast that which had begun to fill. To this evil, buckwheat is subject occasionally all over Northern America. In the State of New York, I obtained a sample of a variety, the excellence of which was said to consist in its not being subject to blight from the midsummer sun.

Leaving Nixon's after breakfast, we were about to enter the forest again at the distance of about a mile, when we were brought up by a burned bridge, the embers of which were still smoking. The fire had communicated to it from the burning woods, and consumed it probably during the night. But we were not long detained. Doffing our coats, and obeying the directions of our companion, Mr Brown, we rigged up beams, along which we guided the carriage across the brook; and then, yoking the horses anew, were under weigh after little more than an hour's agreeable amusement. Twelve miles farther brought us to Steeves's, through narrow clearings on each side of the road, comfortable-looking farms and farm-houses, and occasional good land.

While our horses were resting, I engaged a light waggon, and drove three miles off the main road to the banks of the North River, to inspect an outcrop of limestone at a short distance from which gypsum was said to occur; while, within about a mile, salt springs also were known. We found some good farms along this part of the North River, and good land derived from the mixed calcareous and sandstone debris. The limestone was hard, destitute of apparent fossils, and, as subsequent analyses showed, very pure, and admirably fitted for agricultural purposes. It had been quarried for building, but the application of lime to the land was in this district scarcely known.

The bed of limestone was in contact with, and apparently overlying a coarse red sandstone, which effervesced strongly with acids, and which I afterwards found to contain as much as seventeen and a-half per cent of carbonate of lime, and half a per cent of gypsum. The presence of so large a portion of lime in a sandstone, in the states of carbonate and sulphate, must add very much to the capabilities of any soil that is formed from it.

At Steeves's, where we had stopped to rest our horses, the Petitcodiac ceased to be navigable for canoes, and an Indian portage commenced of twelve or thirteen miles to the navigable waters of the Salmon River, which flows south-westward, and, joining the Kenebecasis, ultimately falls into the River St John. Thirteen miles brought us to Macleod's at the end of the portage, and as many more to Sheck's in the Vale of Sussex, along the bottom of which the Salmon River flows. The country through which the road ran was very parched; the soil sandy and gravelly, as red sandstone soils often are, but mixed with frequent tracts of more useful land. Extensive flats occurred also, upon which, as we passed over them, a wilderness of scrub and other pines prevailed; while the high lands in the distance were covered with the cheering broad-leaved foliage of hardwood trees.

The ridges which on either hand border the Salmon River open out as we approach Sussex Vale, and afford space for a broad valley, into which the traveller looks down as he approaches by the high road, and is at once struck with the scenery as among the finest of its kind anywhere to be seen in the province of New Brunswick. The whole valley is cleared and under culture. It is studded with rounded knolls and ridges of drifted sand and gravel, the wrecks of the sandstones out of which the valley has been scooped. On either hand the ground rises into wooded hills and low mountains, rounded, as such red rocks usually are; while the river flowing through the bottom of the valley, among scattered farms and villages, and churches, and receiving tributaries from every gorge, completes a picture which, when the crops are ripe, as they now were, is very cheering and homelike to look upon in an unsubdued country like this. On the present, as well as on a subsequent occasion, when I visited the Vale of Sussex, my impression was, that I had seen few parts of the province which I should prefer as a place of permanent settlement to the neighbourhood of this beautiful valley.

The air was thick with smoke as we drove down the valley towards our intended quarters; fires were burning in the woods in every direction, and from one spot I counted no less than five-and-twenty burnings visible at the same time in the primeval forest.

30th August.—Early in the morning, I drove over to visit Mr Evanson, an Old Country settler, who owns a fine estate in one of the richest parts of the valley. The intervale land through which the river flows, produces 2½ tons of hay an acre, and is valued at £10 an acre. The undulating slightly elevated land on which Mr Evanson's home farm rests is a red loamy clay, is valued at £7, 10s. an acre, and produces 25 bushels of wheat, 35 of oats, 40 of buckwheat, and 250 to 300 of potatoes. This farm of 400 acres — 200 acres still in wood, 30 in crop, and the rest in grass, with a house for a gentleman to live in, and other buildings—he valued at £3000. This is high for the country, however, being at the rate of £7, 10s. an acre for cleared and uncleared land together.

In this district some farms are let on a money-rent; though, as I have already said, this kind of farming is not popular in North America. It prevails more here, I believe, because this Sussex Vale is an eligible spot, likely to attract by its appearance a settler from the Old Country, and because it is easily accessible from the port of St John, where most of the emigrants land.

I walked over the farm of Andrew Aiton, a Scotchman, which consists of 100 acres of good land of different qualities, lying between the true flat intervale and the upland. It is valued at £7 an acre, and he pays £44 a-year of rent for the whole—or 9s. an acre—which in other parts of the province would be considered a high rent. He had 12 acres in oats, 3½ in buckwheat, 4½ in potatoes, 3½ in turnips, cuts 30 acres for hay, and has the rest in pasture. He keeps 16 cows in milk, 7 young cattle, makes 30 cwt. of cheese, and works his land with one pair of horses. His farm is held on a lease of seven years; he thinks his rent moderate, has no difficulty in paying it, and finds a market at St John. Ayrshire and Durham cattle, he assured me, stood the winter admirably.

Tenant-farming, as I have said, is rare; but the agricultural reader will judge whether landlord or tenant is likely to benefit most by the course of cropping which is followed. On Mr Aiton's farm, some of the oats were the third crop without manure; and eight crops in succession, without manure, are common. Though a Scotch farmer, and keeping so many cattle, Aiton only manured the little patch of turnips he raised. Clover, when sown for the first time, grows three feet high the first year—as it used to do in more virgin days on the Carse of Gowrie in Scotland; but the New Brunswick farmers contrive to take the proud luxuriance out of the land as effectually as any Perthshire farmer, even of the old war times, could do. Red clover and Timothy grass are sown on the wheat. The first year, 2 tons of hay are cut, nearly all clover—the second year, little clover—the third year, all Timothy. After this it is cut eight or ten years, or as long as it yields one-half or three-fourths of a ton per acre. It is then ploughed out for oats, and after potatoes and wheat is laid down again for a similar exhausting treatment by means of hay.

It is evident, therefore, that the tenant-farmer in New Brunswick, if he can only content himself to remain a tenant, has the best of the bargain. He takes the cream off the land and leaves it; and as tenants are in request, he can easily shift to another farm, or can take any good opportunity which may present itself of buying for himself.

An adjoining farm to this of Mr Aiton's is let to an English tenant, for a rent of £90. It consists of 120 acres of rich land, chiefly rich intervale, yielding 2½ to 3 tons of hay an acre, and, with buildings, is valued at £2000. In proportion to the supposed value of the farm, the rent appears lower in this case than in that of Aiton.

A drive of twelve miles down the Salmon River, through similar red land, generally light, but often good, brought us to Baxter's. Within about four miles, two considerable streams join the Salmon River from the west; beyond which the united streams are called the Kenebecasis; and, the river valley narrowing, we leave Sussex Vale behind.

Near the mouth of the lowest of these streams I saw a farm of 500 acres, of which 150 were cleared, including a portion of intervale land. The owner on this farm supports a large family, raises nearly all they require for home consumption, such as grain, flax, &c., and realises from £50 to £150 a-year.

After twelve miles more of a beautiful drive down the Kenebecasis, we arrived at the town of Hampton, where we crossed the river on our way to St John. On leaving Sussex Vale, we found ourselves upon a coarse red sandstone conglomerate, which accompanied us for some miles below Hampton, and formed a surface naturally much more difficult to improve, and therefore less valuable to the purchaser, than the red land we had left behind us. Rocks and stones were strewed plentifully over the slopes. Masses of the hard conglomerate protruded from the soil, and deterred the poorer, less adventurous, or less persevering cultivators. Still, much labour had here and there been expended upon this forbidding surface. Good soils had rewarded those who found courage enough to clear them of trees and stones, and the beautiful landscape gave, in a stranger's eye, a double value to homes established in spite of obstacles in this portion of the county of King's.

At the town of Hampton, I had the pleasure of meeting some of the most intelligent members of the agricultural society of this county. They assured me that an industrious emigrant, without capital, will thrive even in this more stony part of their district; and that, though he must bear privations, yet they never knew a failure. Here, too, the praise of superior industry and perseverance was awarded to the emigrant in comparison with the provincial-born native. This opinion from the mouths of natives is certainly very provoking, since I can sincerely say, after a very long tour in the province, that, in my opinion, a finer-looking body of yeomanry is not to be seen in any part of the world. The first provincial-born generation shoots up tall and handsome men and women, pleasant to look upon. It may be that the more slender form is inclined less to steady labour, and that, with the bodily figure, the habits and tempers of the descendants of industrious settlers change also. But where men are submitted to so many new influences, as they are in this new country, it is very difficult to specify or distinguish how much of any observed change of habits is due to each.

The Kenebecasis, below Hampton, widens into a series of lakes, which, at the distance of about twenty miles, open out into the St John River a few miles above its mouth. Along the borders of this expanded river, intervale and marsh lands occur. Every old upland farm possesses more or less of each of these varieties, which are especially esteemed for the quantity of hay they yield. In wet seasons, when the river is high, the intervale and upland give abundance of fodder; while in dry seasons, when the upland is parched, the water sinks below the level of the marsh lands, which then become accessible, and yield a coarse and rushy, but nutritive hay.

One farm of this kind, now on sale, seven miles below Hampton, consisting of 200 acres in all—of which 30 acres are marsh and intervale, and 80 cleared upland, and which cuts 40 tons of hay—is valued at £600 currency. Another, containing 100 acres of upland, of which about 60 are cleared, and as much marsh as cuts 20 tons of hay, is offered for £400. Men of small capital, therefore, would find at once in this district a comfortable provision, not of luxuries, but of necessaries for their families, by the purchase of such farms as these.

From Hampton to Hammond Bridge the drive was very beautiful. The low marsh and river on our right, the picturesque trap ridges and isolated hills, the rounded conglomerate masses, covered with hardwood timber, and the distant mountain ranges, were all finely brought out and blended together by the sinking sun. The road lay along the upturned edges of conglomerate rocks, dipping at a high angle towards the river; and the surface, as before, was covered with detached masses of the rock, and with drifted sand and gravel. Though the removal of the stones from such land must be very laborious, yet we passed by the way several fine farm-houses, with much completely cleared and apparently good land.

From this point (eleven miles to St John) a change came over the country. We entered a region of more or less metamorphic intermingled slate and trap rocks, on which lay poor, sandy, and gravelly soils. Here and there scattered clearings were seen, but the greater part of the surface is still covered with forests of native pine.

About a mile after we had crossed the Hammond Bridge, I entered the log-hut of a poor Irishman upon a small clearing. Though he had been many years in the province, and was still in the prime of life, he was miserably poor and discontented. He was almost the only one of his countrymen whom I had met with, up to this time, in New Brunswick, who had nothing but complaints to make of the country, and of the hard work he was doomed to. He had landed near the place, and had never moved farther, because *he had no one to depend upon but himself.* He was a kind of squatter, supported himself as a day-labourer, and was evidently one of those inconsiderate, idly-disposed, discontented men, of whom too many, unfortunately for Ireland, love to linger and complain on our own side of the Atlantic.

We reached St John at nine P.M., after travelling three hundred and eighty-two miles with the same pair of horses, in very hot weather, and at the rate of about fifty miles a-day.

1*st Sept.*—In the neighbourhood of St John there is a large tract of dyked marsh, containing upwards of a thousand acres. This consists of a black, spongy, vegetable mould, very different from, and inferior in quality to, the rich, alluvial, dyked land of the head-waters of the Bay of Fundy, hereafter to be described. At the mouths of nearly all the rivers which empty themselves into the Atlantic, along the shores of New England and New Brunswick, such inferior marsh lands are met with. They abound in vegetable matter, yield large crops of hay in favourable seasons; but when dyked round and defended from the sea, rarely form rich and productive corn-bearing land. Like our peaty or fenny soils at home, they are propitious to green crops and the growth of straw, but do not generally fill the ear.

The marsh land of St John lies in a narrow valley, bordered by high ground on each side, but itself very little elevated above the sea. The upper end of the flat is only two feet above high-water mark; but as the tide rises here twenty-seven feet, its height is considerable above mean-water level, and the entrance of high tides is prevented by a sluice at the mouth of the valley.

I visited what is considered one of the best farms on this flat. It consists of 120 acres of marsh, and 100 of upland. The upland is partially cleared, and affords pasture and firewood, but the marsh alone is under arable culture. The whole is rented for £150 a-year currency. It requires high manuring; but when well cultivated, any part of it, the tenant said, would produce four tons, and I was assured that five tons of hay was occasionally reaped from such land. About St John, and generally along the Bay of Fundy, fogs occasionally prevail. St John has about a month of summer fogs, which are not considered prejudicial to health; and though unfavourable to Indian corn, they encourage the growth of grass. Turnip culture has extended very much during the last three or four years in the province; and upon this dyked land farm, and on those of Dr Peters and Mr Davidson, both within a mile or two of St John, I saw excellent crops of this root.

That of Dr Peters is an upland farm, the soil a light granitic sand, and gravel, naturally poor, and exhausted by previous tenants. The practice of tenants hitherto, in this neighbourhood, had been the same in principle as I had already seen in so many other places—to take a crop of oats, lay down to grass, and cut as long as anything was to be got, then to try oats again, followed by grass as before; and when to such alternations it would yield nothing more, to give it up to the landlord. It contains 160 acres, of which 50 were cleared when he bought it for £320, or at the rate of £10 an acre all over. By manuring heavily with cowfeeders' and slaughter-house manure, which is here got for sixpence sterling a cart, he is compelling the exhausted land to yield beautiful crops.

The adjoining farms, containing 120 acres of upland, similar to that of Dr Peters, and 40 acres of marsh, capable of growing ninety tons of hay, with house and barns, was on sale at the price of £600.

These farms are near a good market, but they are on an old slate country, distinguished for thin, cold, poor, and gravelly soils, over which, with the exception of a few more favoured spots, only scattered clearings and a sparse population extend nearly all the way along the northern shores of the Bay of Fundy, from St John to St Andrews, a distance of sixty-five miles. It is repulsive, therefore, to the emigrant, and will be occupied and tilled much more slowly than some of those inland parts of the province of which I have already spoken.

Dr Peters had two farm-servants, to each of whom he paid £40 a-year without board. Mr Macarthy, on the dyked land, gave his head-man £30 currency in money, with a house, fuel, potatoes, and a cow; and to his other men from £15 to £25, besides their board. But in St John, and the adjoining counties, wages are lower than in any other part of the province, because so many (especially of the Irish) emigrants linger here—as they do elsewhere—near the port they arrive at, and work for smaller wages rather than proceed into the interior, where better land and better wages are to be obtained.

The following table exhibits the average rate of wages for agricultural labour in the several counties of the province, in addition to board, washing, and lodging, in sterling money :—

Counties.	By the Day.						By the Month.						For the whole Year.		
	Summer.			Hay-time and Harvest.			Summer.			Hay-time and Harvest.					
	£	s.	d.	£	s.	d.	£	s.	d.	£	s.	d.	£	s.	d.
Saint John,		...			...		2	0	0		...		19	10	0
Charlotte,	0	2	3	0	2	8	2	7	8	3	5	0	22	0	0
Westmoreland,	0	1	9	0	3	0	1	12	6	3	5	0	20	13	9
King's,	0	2	6	0	4	3	1	15	0	3	6	8	18	12	6
Queen's,		...			...		2	3	4	3	3	4	18	17	6
Sunbury,		...			...		1	10	0	2	10	0	22	3	4
York,	0	1	10	0	3	0	2	3	4	3	0	4	24	0	0
Castleton,		...			...		2	0	0	3	0	0	25	0	0
Albert,		...			...		2	5	0		...		25	0	0
Kent,		...			...			...			...		25	0	0
Northumberland,		...			...		2	10	0	3	0	0	27	0	0

The wages at the present time in Berwickshire (Scotland) for an able-bodied man, who can manage a pair of horses, is £14 to £16 a-year, with board and lodging. The wages in New Brunswick, therefore, even in the depressed state of home agriculture, are not so much greater as to induce the farm-labourer, who can obtain employment, to emigrate to this colony for the purpose of working as a labourer. But the prospective advantage to him is, that he can obtain land for himself at so cheap a rate, when—after working a year or two for wages, (which I would recommend him contentedly to do,) till he has obtained so much acquaintance with the country and its customs as to enable him judiciously to select a location for himself—he has saved money enough to pay the first instalment for his land, and keep his family during the first winter season.

CHAPTER V.

By steamboat from St John to Portland, in Maine, and thence by railway to Newhaven, in Connecticut.—Alleged rudeness of American manners.—Country houses along the approach to Boston.—Drought in New England.—Farming in Connecticut and Massachusetts.—Diffusion of agricultural periodicals.—Cider-making in Connecticut.—Yale College.—Number of Students.—Expense of residence.—Inferior position of professional men.—Salaries of clergymen.—Estimation of lawyers and medical men.—Favouring of quacks in the majority of the States.—Medical schools in the United States.—Opening for European medical practitioners.—Elm-trees of Newhaven.—Tree-toad.—Fairhaven; its oyster-trade.—Two species of American oysters of large size.—Consumption of oysters in Massachusetts.—Railway to Albany up the Housatonic valley.—Light soils and Indian corn of this valley.—Post-tertiary clays and sands of the upper valley of the Hudson River and of Lake Champlain.—Natural forests which grow upon them.—Power of their soils to resist drought.—Exhalation from the leaves of plants.—Relation of the porosity of a soil to this exhalation.—Why drained and mellowed clay is moister in a hot summer than undrained.—Schenectady.—Valley of the Mohawk.—Character of its soils and produce.—Rich bottoms of the Mohawk resting on the Utica slate.—Broom corn, (*Sorghum saccharatum,*) its extensive cultivation in this valley.—German flats.—Utica a thriving manufacturing town.—German population.—Change in the meaning of familiar words.—Importance of keeping the English language pure.—Choice of judges by popular election.—Apparent danger of this practice.—Titular judges and generals.—Popped corn.—Structure of Indian corn.—Extraction of oil from this grain in the Western States.—Flour of Indian corn; varieties in its colour; used in the adulteration of wheaten flour.—Digestive powers of animals.—City of Rome.—Mr Clay.—Verona.—Change in the character of the country.—Arrival at Syracuse.

Sept. 4.—At eight in the morning, I went on board the small steamer which plied between St John and the town of Eastport, in Maine, on my way to the state of New York. Steaming in the Bay of Fundy is not always agreeable, even in large boats. The day was fine, however, and we made the distance of seventy miles to Eastport in about eight hours. There we were received on board of a larger steamer, which conveyed us to Portland, in Maine, by half-past eight on the following morning, in time for a railway train about to start for Boston. At 2 P.M. I arrived in Boston, being five and a half hours for a hundred and eleven miles, five hours being the usual time. Starting again at 4 P.M. from Boston by the New York line, I reached Newhaven, in Connecticut, at 11 P.M., being a hundred and sixty miles in seven hours, or about twenty-three miles an hour.

In this rapid run through New England, only three things made a permanent impression on my mind. These were, first, that the general rudeness of the people which travellers speak of is not perceptible in New England generally. It may be more striking in the Western States; but if, on our home railways, all classes were indiscriminately mixed up in large carriages—cars, as they call them here—containing fifty or sixty people, I doubt if Old England passengers would, as a whole, behave as well as those of New England do. The second thing was the numerous country boxes or cottages, of all fashions and sizes, with their white painted walls and green jalousies, which skirted the railway during the last twenty miles of our ride to Boston. This is a peculiarly English feature, and indicates the existence among our Transatlantic kindred of that love of green fields, and of a quiet country life, which characterises so much our island-home. By the operation of this feeling, as is the case around our own great cities, the wealth of the growing commercial city of Boston is carried out to the country residences of its merchants, and is on a thousand spots in course of being expended in clearing and improving the stony, and in fertilising the gravelly and sandy soils of which a large portion of the surface of Massachusetts consists. And my third observation was, that though the drought of Nova Scotia and New Brunswick had extended into Maine, its effects became less perceptible as I advanced westward into the other New England States, till, in Connecticut, the fields looked as beautifully green as I had seen them last at the mouth of the Mersey; and the after-grass was abundant.

Along the Connecticut River, which runs through the centre of this state, there is much good land, and tobacco in considerable quantity is grown upon it. The produce per acre is from 1500 to 2500 pounds of marketable tobacco. This is a very exhausting crop, as all *leaf* crops are; and land must be generously used which is to continue long to yield crops such as these.

The farming of Connecticut is said to have greatly improved during the last fifteen years, and this is ascribed in part, and probably with some truth, to the extensive circulation and perusal of agricultural papers. In the small country town of Farmington, (of two thousand inhabitants,) for example, a friend of mine assured me that not less that fifty agricultural papers were taken in by the inhabitants. And most of these papers—the *American Agriculturist* and *American Cultivator*, for example—are really well and usefully got up, and filled with valuable information.

A similar improving character is ascribed to the farming of Massachusetts, but less is said in favour of New Hampshire and other parts of New England. Of the first settlers in Connecticut and Massachusetts, many were from the west and south-west of England, from which places they naturally brought some parts of their old home husbandry, as that of apple-growing and cider-making. The quantity of cider formerly made and drunk in this state is said to have been immense. It is, I suppose, from this being a staple branch of husbandry, that our home saying, "Great cry and little wool," originating in the noise made at the scraping of a pig, has here assumed the form of "More cry than cider."

9th Sept.—Newhaven in Connecticut is known in Europe chiefly as the seat of Yale College, one of the oldest and most respected of the academical institutions in the United States. It was founded in 1700, has at present 531 resident students, of whom 386 are under-graduates, 52 students in its theological, 33 in its law, and 41 in its medical school.* The Orthodox Congregational body are the prevailing denomination in the state of Connecticut; and Yale College, though in no way exclusive, and having no tests, is under the management of trustees and instructors, who are for the most part of this denomination. Rooms are provided in the college buildings for about one-half of the under-graduates, the rest living in lodgings in the town, and all dining in private houses or clubs, as is the custom in the German universities. The total annual expense of a residence at college is about 150 dollars, or a little more than thirty pounds, made up as follows,—

Tuition fees, ...	33	dollars.
Room-rent, &c., ...	21	—
Board (forty weeks,) . .	64 to 90	—
Wood, lights, washing, furniture, stationery, &c., . . .	32 to 71	—
Total, ...	150 to 215	—

So that about 200 dollars, or £42 sterling, will, on a liberal scale, defray all the necessary residence expenses of under-graduates. The medical institution in connec-

* The above was in session '49-50. In the present of '50-51, the number of under-graduates is 432, and the whole number of resident students, 555.

tion with Yale College was established in 1810, and the theological department as late as 1822. The former has six and the latter four professors. I do not know whether the expense of the purely professional education is greater than that of students in arts. A step has recently been taken towards a provision of special education for the agricultural and higher industrial classes, by the establishment of chairs of chemistry in its application to the arts, and of chemistry in its relations to agriculture and physiology. These departments have been placed respectively under the charge of Benjamin Silliman, jun., whose father has so long enjoyed a European reputation, and, by his writings, made his college known where otherwise it would never have been heard of; and of my friend, and former pupil, Professor Norton, who is already favourably known in this country, as well as in his own.

A circumstance which early strikes the European traveller in the United States, is the comparatively small consideration in which professional men are held, and the small salaries they in general receive. The former may be supposed to arise from the more universal diffusion of a certain amount of instruction than is the case in most European countries—and this is no doubt in part the cause. But it is partly due also to the theoretical and practical political equality of all citizens, which appears to induce, among the masses of ordinarily educated men, an impression that higher intellectual gifts or attainments are, generally speaking, no sufficient reasons for social distinctions or higher consideration. Every man you meet thinks himself capable of giving an opinion upon questions of the most difficult kind; and, for the most part, the masses seem, by their choice at public elections, to prefer to be guided by the less rather than by the more educated of their fellow-citizens.

In the country districts, five to eight hundred dollars

are considered a fair salary for a clergyman. In the cities, from eight to twelve hundred are given, and, in rare cases, or to especial favourites, fifteen hundred.*

Of professional men, the lawyers succeed best, as the same theoretical equality which makes a man think his own opinion as good as his neighbour's naturally promotes litigation, and makes the lawyer necessary, and the clever lawyer sought after and honoured.

Medical men are perhaps the worst off, as, in most of the States, the educated physician has no defence against the quack. In only six out of the thirty-three States are there any laws making licenses necessary to the practice of medicine, or which place the educated physician in any respect in a better position than the pretender: these are New Jersey, Delaware, the district of Columbia, Georgia, Louisiana, and Michigan. In the other States, any man may call himself a doctor, may practise, and may sue for his fees; and, in many cases, a discerning public prefers the self-taught genius to the man of education.

In the United States there are at present thirty-five medical schools, with 4566 students, which send out on an average about 1300 graduates annually; but it is calculated that, to supply fully the demand of the growing country, 2500 graduates should leave these schools every year. Under such circumstances, quackery must abound, unless the schools of Europe aid largely in adding to the stock of home-made surgeons and physicians.

The salaries paid to the clergy may be taken as a fair measure of their status among a Protestant people. Hence, if £100 to £160 a-year be the usual stipends paid to the clergy as a body, in a country where the labouring man and the mechanic obtains a considerably higher wage than with us, the estimation in which he is held

* A hundred dollars are about £21 sterling.

must also be less elevated above that of the working classes than it is in the Old Country. But things will probably alter as age creeps upon the young States, and as men of leisure arise, who shall really have time, not merely to skim the surface of the various departments of knowledge, but to sound their depths, and bring up from below some of those rarer pearls, which long intellectual labour alone secures, but which even unintellectual men can admire, and are willing both to praise and purchase.

Newhaven stands on a broad sandy and gravelly flat, which extends along the head of one of those numerous bays which, at short distances, indent nearly the whole of the Atlantic border of the New England States. The town is regular, well built, and increasing, but presents nothing so worthy of remark as the numerous rows of elm-trees with which its public square and some of the more private streets are ornamented. Of the beauty of the American elm I have already spoken; but though it appears to thrive well here, I saw none possessed of that rare beauty which has so often struck me in single specimens, groups, or rows of these trees in their native forests, or by the sides of streams, where art had not yet interfered to alter the arrangements of nature.

A novelty which early arrested my attention at Newhaven was the chirping or whistling of the tree-toad, *Hyla versicolor*, which, from the noise one hears filling the whole air as evening approaches, must live in great numbers among the branches of the elm-trees. Of toads possessing a peculiar prehensile apparatus resembling claws, which enables them to cling to the under side of smooth branches, which devour insects, and are usually found upon trees, there are at least two species in the United States—the northern tree-toad and the squirrel tree-toad. The former is the only one generally known in New England and New York. It is about two inches in length, is broad across the head and shoulders, has an irregular dusky cross on its back, and varies in colour at will, from grey to green. This variation in colour makes it difficult to be seen upon trees. During damp weather it is peculiarly clamorous, especially in the fall of the year. It is here considered, along with the fall of the leaf, as one of the sensible harbingers of winter. As evening approached, these animals began to chirp in the trees opposite my window; and from that time till I fell asleep in bed, the air was filled with a low whistling or whispering sound, somewhat resembling a sharp rustling of the wind among the leaves. The tree-toad is rarely heard in the morning, when a perfect stillness prevails. It is difficult to convey an exact idea of the sound this animal emits, as it is something of a ventriloquist, and often deceives the most attentive observers, who are looking for and endeavouring to catch it. The southern or squirrel tree-toad is brown in colour, and has a length of only an inch and a quarter. I have not heard its note, if it has any.

Opposite to Newhaven, on the other side of the bay, stands Fairhaven, a town with three conspicuous churches, which is supported entirely by its trade in oysters. All along this coast, from the Gulf of St Lawrence to the Delaware and the Chesapeake, oysters abound, either naturally or in consequence of artificial importation, and are an article of extensive consumption and large traffic. They are a constant dish at table, sometimes as often as twice a-day. They are shelled, and, with the liquor contained in the shells, are sold at about a dollar a gallon. In winter, these shelled oysters, frozen in their own juice, are sent by the railways in ton loads, and are greatly prized in the interior.

It would appear that the oyster abounded in most of the New England bays when the country was first colonised; but the supply for the markets of all the large towns is now obtained from the south,*—from the bays of New York chiefly, and from the Delaware and the Chesapeake. Two species of oyster are known on this coast—the *Ostrea borealis* and the *Ostrea virginiana*, commonly distinguished as the northern and southern oysters. Both grow to a large size compared with our English and Scotch oysters. A common size of shell in the northern species is five and six inches, though some arrive at twelve inches in length and six in breadth: the southern shell often measures twelve to fifteen inches in length, by not more than three inches in breadth.

The fishermen of New England and New York import the oysters from their native beds, on the mud banks of the southern shores, in a young state, and plant them in layers near their own homes, where they increase in size and flavour, and grow in numbers, though not fast enough to supply the home demand. I had no opportunity of obtaining accurate information in regard to the oyster-trade which supports the town of Fairhaven; but its extent may in some measure be judged of by what Dr Augustus A. Gould has stated in regard to the same trade in Boston. The oysters consumed in Massachusetts, he says, amount to about 100,000 bushels a-year, and they are sold wholesale at a dollar a bushel. The beds—or *scalps*, as they are called in Scotland—are chiefly situated at Wellfleet, near Cape Cod. About 40,000 bushels of small young oysters are annually imported, and planted in this place. Thirty vessels, of about forty tons each, are engaged in the trade, employing one hundred and twenty men for three months in the year. In the autumn the oysters are taken up again, and those are selected which have grown to a proper size for market. The vast quantities of oyster-shells one sees employed as rubbish to fill up hollows in the outskirts of

* Dr Gould, *Invertebrata of Massachusetts*, p. 358.

Boston, are proof enough to a stranger of the vast consumption of this fish which must there take place.

In the bay of Newhaven there are certain public layers or beds, which are open to every comer at certain prescribed seasons of the year; but there are also numerous private layers, which are the property of individuals, whose rights are strictly preserved by law. They form, in fact, their stock in trade.

10*th Sept.*—After spending four days at Newhaven, I started this morning early for Syracuse, in Western New York, in company with Professor Norton. The annual fair or show of the New York State Agricultural Society was to be held at that place in a few days, and I had, at the request of the Council of the Society, agreed to deliver the Annual address. For the first eighteen miles we skirted Long Island Sound, on the New York line. At Bridgeport we took the Housatonic line, which runs inland along the river of the same name for ninety-seven miles, when we joined the western line, which carried us to Albany, a further distance of thirty-eight miles. The train was heavy, the number of passengers large, and the delay caused by the luggage considerable; so that we did not arrive at our hotel in Albany till about eight o'clock in the evening. In this city there are several good hotels. We went to Delavan's, which, like several of the others, is a temperance house, and is particularly convenient for railway travellers. About a hundred and fifty of our fellow-passengers by the railway came to the same house, and were all accommodated. About an hour after our arrival, tea was announced, and we all sat down together in the dining-hall. On the following morning, all who were going west were called an hour and a quarter before the train started; in half-an-hour after this, breakfast was on the table; in another quarter, carts were at the door for the luggage; and at six A. M. we were again steaming from Albany, along the railway, towards Western New York.

The day was fine as we came up the Housatonic Valley. Soon after leaving Newhaven, we passed from the new red sandstone of Connecticut to the old slate and metamorphic rocks, and along these we continued most of the way to Albany. The soil in general was light and stony, resting for the most part on gravelly and sandy drift. Such light soils are adapted to the culture of Indian corn, and we saw many fields of this grain as we came along. It was indeed the only breadstuff we saw during the day, with the exception of an occasional field of buckwheat, where the soil was especially poor. The rolling upper valley of the Hudson River, towards which we descend on approaching Albany, contains much strong yellow clay, the same post-tertiary clay which borders Lake Champlain, and thence stretches north and east along the banks of the St Lawrence.

This formation is a hundred feet thick upon Lake Champlain, and it forms high banks along the Hudson River in the neighbourhood of Albany. It consists, in the under part, of stiff blue clay, resting upon the grooved rocks or upon beds of drift; above this, of a lighter coloured clay, pale brown or drab, containing shells of existing species; and, over all, a deposit of yellow sand, sometimes loamy and fertile, but often barren, and sustaining only stunted pines. The soils, the natural vegetation, and the agricultural capabilities of the district over which this formation extends, vary according as the upper sand remains on the surface, or has been more or less completely removed by natural causes. The stiff clays bear natural forests of hardwood trees of various kinds; but when cleared, and put under crop, the clay hardens, cracks, and becomes parched under the hot suns of summer. The sandy loams which rest upon the clay nourish numerous pines stretching along the country as broad pine-barrens, on which the beautiful white pine is the predominating and characteristic tree; while the pure sands, which form the uppermost layer, are covered by the yellow pine (*Pinus mitis.*) Of these latter, when brought into cultivation, experience has shown that the loamy sands are less affected by the heat than the stiff clays, but that the apparently purer sands bear the drought better than either. The mechanical porosity of a soil, in fact, has almost as much to do with its power of resisting scorching heats in summer, as any other of its properties, either physical or chemical.

This will appear very plain by a reference to one consideration only. The leaves of plants exhale watery vapour in large quantity, and this quantity increases with the intensity of the summer heat. No experiments have been made on this subject in America, I believe; but in the cooler climate of England, an acre of our cultivated crops has been estimated to exhale from its leaves, during the four months of summer, about three million pounds of water;* while in the same time there falls, of available rain, not more than eight hundred thousand pounds—little more than a fourth of the exhalation. The soil and the plant, therefore, must inhale from the air during the cool night far more water than they derive from the rain that falls, however grateful and refreshing to plants this may be.

Now, the more porous a given soil is made, the more absorbent it is of moisture from the air, and hence one reason why light soils suffer less from drought than heavy soils—why heavy clay soils are so much improved

* Mr Lawes, in some recently published experiments, found the quantity of water exhaled to be less than it was estimated to be from the experiments referred to in the text. But his results also show that plants must derive moisture from some other source than the rain-water that sinks into the soil.

by under drainage, which prevents the sun from baking, hardening, and making them compact and non-absorbent—why drained land is in reality moister in summer, because more porous than the undrained—and why even in our tropical, sun-burned, sugar colonies, thorough-drainage has been found to promote the growth of the cane on stiff clay soils, and to increase the yield of sugar.

11*th Sept.*—On leaving Albany, we climbed a steep gradient of one in eighty, till we reached the upper level of the Hudson Valley. We then crossed about sixteen miles of a rolling country—as a surface of rounded hills and ridges succeeding each other is here graphically called—of pale yellow post-tertiary sand, to Schenectady, a town of eight thousand inhabitants, situated in the valley of the Mohawk. This sandy tract forms a part of the pine-barren of which I have spoken, as overlying the post-tertiary clays of the Champlain and Hudson River valleys.

The Mohawk is a river of considerable size, which falls into the Hudson from the west, nearly opposite to the city of Troy, and about eight miles above Albany. The valley through which it flows, and which the railway follows, occupies a distinguished place in the early history of the State of New York. It was the site of many of the earliest settlements in this western country, and the scene of many interesting passages with the native Indians. It is still the first fine valley which attracts the longing regards of the European emigrants, whom the Erie canal and the railroad carry in thousands along the banks of the Mohawk River towards new homes in a still farther west.

As we rushed up the left bank of the river, the Erie canal accompanied us on a higher level on the right or opposite bank. It was crowded with vessels going and coming, some laden with piles of casks filled already with the flour of the present season. On the uplands and slopes, sand and gravel prevailed, and there buckwheat and Indian corn were cultivated. The low alluvial flats or intervales, of which many occurred in the bottom of the valley, bore heavy crops of Indian corn, (maize,) and of broom corn—a crop unknown to our home farmers.

The valley of the Mohawk, when compared with the more western parts of the State, is distinguished agriculturally by its production of these two kinds of corn; while the western counties, from the Oswego River to the Genesee River, and onwards even to Buffalo, form the wheat region which, under the name of the Genesee country, attracted so much attention as an eligible place of settlement towards the close of the last century.

Sixteen miles above Schenectady we passed the town of Amsterdam, above which the Mohawk River runs for a great many miles through the Utica slate. This slate, an upper part of the lower Silurian rocks, is calcareous, crumbles very readily under the influence of the weather, and forms a rich, clayey, yet free and open soil. It is when flowing over this formation that the Mohawk Valley expands into extensive flats of great beauty, now on this side of the river and now on that, where apple orchards abound, and fields of Indian and broom corn cover the rich alluvial soils.

Of these two crops, the broom corn (*Sorghum saccharatum*) was the most abundant for ten miles above and below the Little Falls of the Mohawk. To the eye of a stranger, who sees it for the first time, it resembles a crop of Indian corn; it is nearly of the same height and strength of stem, but it has a narrower leaf and a branching head. This head or top is the only part which is collected. When the fruit first appears, this long head, which bears the seed, is bent down at a right angle about two, or two and a half, feet from the top, so as to break the stem and arrest the flow of the sap. This upper part is then left a few days to dry, when it is cut off with a knife, leaving a high stubble of one or two feet, which is afterwards ploughed in. The seed is beat out of the tops, which are trimmed and sorted, and are then ready for sale or for manufacture into brooms on the spot. Brooms of this material are of almost universal use in North America, and in favourable localities it generally proves a profitable crop.

Above Little Falls, where thriving woollen mills are moved by the water power, Indian corn alone, ripe, cut, and standing in sheaves, covered the cultivated fields. Such fields, along with extensive meadows, on which large droves of milk-cows cropped the aftermath, filled the entire bottom of the valley. About Herkimer, which is eighty miles from Albany, the flat lands expand into a beautiful broad valley, called the German Flats, on which the quantity of Indian corn raised appeared quite immense.

The name German Flats indicates the native home of the settlers who first colonised this part of the Mohawk Valley. Indeed, places with German names abound in this quarter. Along the river lie the townships of Minden, Oppenheim, Manheim, Danube, and Frankfort; and the mixed English and German sign-boards in the villages, show that there are numbers of the inhabitants to whom German is still the more familiar tongue.

The city of Utica, fifteen miles farther, stands on the right bank of the river, in the midst of a broad expansion of the valley, resembling the bed of a great lake. It is a clean thriving place, of about fifteen thousand inhabitants, has a striking main street—is the seat of flourishing manufactories—is distinguished especially for its woollen mills, but is adding rapidly also to the number of its cotton factories. It is admirably situated for the trade of the Western States. That it is surrounded by a large German population is shown by the Gasthaus, Buchhandlung, and other German announcements which are to be read in abundance on the fronts of the houses, and over the shop-windows.

Winter rye appears to be sown very early in this State. I had seen a field already sown upon the sandy pine plain between Albany and Schenectady; but near Herkimer, on the banks of the Mohawk, I passed one already above ground, (11*th Sept.*) and forming a beautiful green braird.

Fourteen miles, chiefly through sandy, gravelly, and swampy barrens, brought us to Rome, an aspiring village in a town(ship) of the same name—remarkable chiefly for its large hotels, and for the outlines of fine streets which are hereafter to be built up.

Among the new meanings attached to old words which one observes in this State of New York, is the sense in which the word town is used even in legal phraseology. For brevity's sake, I suppose, the word township has gone out of use, and town has taken its place. A county is divided into so many towns; and in a town, so many villages, or even a city may spring up. This change in the meaning of a familiar word perplexes very much the stranger, is productive of inconvenience here, and is much more injurious to the maintenance of a pure common speech between the Old Country and the New, than any introduction of absolutely new words can ever be.

Thus, a town in western New York may be a square of land, without a house or an inhabitant upon it; and in some places a street is merely a road through a partially cleared township. Then all collections of houses are necessarily called either villages or cities. Where places are incorporated, they become corporate villages; and if the aspirations of the inhabitants make them discontented with this inferior title, they make influence with the legislature, and, by means of a little *log-rolling*,* get themselves separated from the town, and raised at once to the dignity of a city. The introduction of new words enriches a language, but to give an old word a new signification not only perplexes the hearer, but makes the language poorer.

In a township, a village springs up which naturally assumes the same name—in the town of Pompey or Cæsar, the villages of Pompey or Cæsar spring up. Then a second village, a third, a fourth, a fifth, and these are called respectively, Pompey East, Pompey West, North, and South. And, finally, at the cross roads which lead to these several villages another springs up, and this becomes *Pompey four Corners*, and all these in the *town* of Pompey.

Of course, the attaching of a new meaning to a few

* In Bartlett's *Dictionary of Americanisms*, we have the following explanation of *Log-rolling* :—" In the lumber regions of Maine, it is customary for men of different logging camps to appoint days for helping each other in rolling the logs to the river, after they are felled and trimmed—this rolling being about the hardest work incident to the business. Thus the men of three or four camps will unite, say on Monday, to roll for the camp No. 1—on Tuesday for camp No. 2—on Wednesday for camp No. 3—and so on through the whole number of camps within convenient distance of each other.

" The term has been adopted in legislation to signify a like system of mutual co-operation. For instance, a member from St Lawrence has a pet bill for a plank road which he wants pushed through ; he accordingly makes a bargain with a member from Onondaga who is coaxing along a charter for a bank, by which St Lawrence agrees to vote for Onondaga's bank, provided Onondaga will vote in turn for St Lawrence's plank road.

" This is legislative *log-rolling ;* and there is abundance of it carried on at Albany every winter.

" Generally speaking, the subject of the log-rolling is some merely local project, interesting only to the people of a certain district ; but sometimes there is party log-rolling, where the Whigs, for instance, will come to an understanding with the Democrats, that the former shall not oppose a certain Democratic measure merely on party grounds, provided the Democrats will be equally tender to some Whig measure in return. J. INMAN."

single words is of no great consequence in itself; but I mention this instance for the purpose of illustrating the way in which changes may insensibly be induced in our mother tongue, while the words in use remain the same. Where two countries speaking the same language have different literary and courtly centres, the gradual growth of differences in the speech can scarcely be avoided. Such causes of diversity between the older forms of speech which prevailed at the same period in London and Edinburgh anciently existed in the rival courts of those cities; diversities of a similar origin have given rise to the now distinct dialects, almost distinct languages, of Sweden, Denmark, and Iceland; and if the German tongue remains still common and intelligible over the whole Fatherland, it is because a common literature still binds the people of the various governments together, and a common theatre and drama mould the speech of Vienna, Berlin, and Hanover, in spite of the peculiarities of fashion which give a special character to the pronunciation and choice of words in each of the several courts.

No language, probably, ever existed which it is of so much importance to mankind to preserve pure as the English. It is now spread over so vast a portion of the civilised globe that in it we may almost hope to see realised, for the intercommunion of the people, what the Latin tongue was to the learned of former centuries. It is, emphatically, the language of constitutional freedom and of civilised progress. If the community of speech puts within the reach of the most ignorant and ill-intentioned editor of the meanest periodical, in either country, the foolish or ill-natured sayings of his brethren of the craft in England, America, Africa, or Australia, and thus enables bad men to excite evil passions, to create enmities, and to awaken national bitterness, which, for the welfare of the Anglo-Saxon race and the general good of mankind, it is desirable rather to soothe down and suppress—it forms a bond of union also which makes the intellect of this vast population, and all its triumphs, the almost instantaneous property of its remotest members; which makes countless hearts beat, and pulses throb in common sympathy, and which opens no less readily to the accents of peace than to those of war an entrance to many ears—as the sound of the ancient tongue of his country is said to command an immediate and easy admission for the peace-bearing doctrines of the gospel to the heart of the Irish mountaineer.

Among the advertisements contained in the New York papers of to-day, I was struck with one which referred to the elections to be held in that city on the first Tuesday of November, and which specified, among other officers to be elected by the suffrage of "all male citizens of twenty-one years of age,"—

A Justice of the Supreme Court.
A Judge of the Court of Common Pleas.
A Judge of the Superior Court.

This practice of electing Judges by popular suffrage, and for a limited number of years, appeared to me, with my home opinions and historical recollections, one of the most suspicious of the novelties which have lately been introduced into the New England States.

This practice is of recent origin in the State of New York. The constitution of this State, as amended by the State Convention of 1846, enacted that all Judges should be chosen by the suffrages of the electors of the district, great or small, over which their functions were to be exercised, and that their tenure of office should not exceed eight years. Justices of the Peace are chosen by the electors of the several town(ships,) at their annual town meetings, and hold office for four years; and they are paid a *per diem* allowance for their services in sessions, and fees for other services. The Judges of the

Supreme Court are chosen by the electors of the Judicial district over which they preside, and the Judges of the Court of Appeals by the electors of the whole State. These Judges have fixed salaries, are elected for eight years—going out by rotation—and may be removed by a concurrent resolution of two-thirds of the Assembly and of a majority of the Senate of the State.

There are, in all, thirty-six Judges of the Supreme Courts. Four sit in the Court of Appeals, and four in each of the Supreme Courts located in the eight judicial districts of the State. The professional rank of all is equal, and their salaries alike—2500 dollars a-year. This salary for an uncertain period only, and the liability to be thrown out of office without a retiring allowance, at the popular will, will scarcely enable the bench of the State to secure the highest talent of the bar, in a litigation-loving country like this.

To those, also, who are as sensitive in regard to the purity and independence of the Judges of their country, as to the perfection of its political institutions—especially to persons born in Europe, or who are familiar with the history even of English courts of justice—this mode of election by popular suffrage appears by no means void of danger. It was deservedly considered a great triumph when the appointment of Judges for life liberated the English bench from the influence of the Crown, and when public opinion became strong enough to enforce the selection of the most learned in the law for the highest judicial offices. Now, passing over the objection which some will strongly urge, that the popular electors are not the best judges of the qualifications of those who aspire to the bench, and that the most popular legal demagogue may expect to obtain from them the highest official appointment,* it may be reasonably asked,

* In the estimation of the *Barnburner* section of the democratic party, this objection has no weight—one of their principles being "to

whether popular influence in seasons of excitement, and upon questions of great moment, may not bias the minds of Judges whose appointment is in the hands of the people—whether the fear of a coming election may not deter them from unpopular decisions. The influence of a popular majority may here as profoundly pollute the fountains of justice as the influence of the Crown ever did among us at home. It is not wise to expose good men to such temptations.

The practice is consistent with the theory that all power is in the people, and that all patronage ought to be exercised by them. And it may be that ignorance and passion may not more frequently interfere with the appointments of the people, than favouritism and political influence do with those made by sovereigns and their ministers. At all events, it is said that hitherto the system has worked well, and that good selections have been made; though I have heard it distinctly acknowledged that certain Judges had been excluded at recent elections in this State of New York, because of their known opinions in reference to important public questions which were likely to come before the courts.

As there are no retiring pensions, the want of success at a re-election necessarily sends a man back to the Bar. And as in Scotland "ance a bailie, aye a bailie" is the rule, so once a judge or general, always a judge or general, in the United States. Hence the numerous titular judges to be met with among the lawyers, and of generals and colonels among all classes of the people.

Popped corn is a novelty which the European travelling

fill all offices with men taken from among the people, and not to confine them to those who live by office and make politics a trade." Those who are themselves fit to be taken from the counter or the plough to fill the highest political offices, cannot be unfit merely to select those who are to fill the highest judicial offices.

towards the west will, in this maize-growing region, probably meet with for the first time. Boys with baskets attend upon the cars at the stopping-places, selling it at so much a quart. Indian corn, beneath a thin but double epidermis, consists of a semi-transparent, hard, horny, or flinty part of a yellow colour; within this, of a white, soft, opaque, starchy part; and within all, at the base of the seed, of the germ or *chit*, as it is called in America. The horny part consists of starch, of oil, which under the microscope can be seen in globules, and of a proportion of vegetable albumen.* The soft white opaque part is chiefly starch, and the chit almost entirely albumen. This is prettily shown by cutting off with a knife the outer part of a grain of Indian corn, and applying to the pared surface a drop of a solution of iodine: the soft white part will become entirely deep blue; the horny part blue only in streaks or patches, while the chit will be scarcely affected. The substance thus turned blue is starch. If another grain of corn be touched in the same manner with a solution of verdigris, (acetate of copper,) or of blue vitriol, (sulphate of copper,) only the chit will be coloured blue—showing that, as a whole, it differs entirely from either of the other parts.

But the cutting of the seed across, and especially if the cut surface be coloured by the application of any of these solutions, shows that the relative size and position of these several parts varies very much with the variety of corn we examine. In some the horny part is large, as in the varieties known by the names of brown, Canada, rice, and pop corns; while in others, as in the flat southern and in the Tuscarora, the white starchy part predominates. In some, as in the pop and Canada corn, the horny part entirely surrounds the soft starchy portion; in others, as in the flat southern, it forms an

* Nearly the same thing as the white of egg.

irregular layer round the side only of the seed; while in others again, as in the rice-corn, the seed consists almost entirely of a large horny part and a large chit.

As the oil exists in this horny part, it is obvious that those varieties in which this part is large ought generally to contain most oil; and such is found to be the case. It was an ignorance of these differences among the varieties of Indian corn which a few years ago caused the violent dispute between Liebig and Dumas, or rather the hasty contradiction by Liebig of the statement of Dumas, as to the large quantity of oil contained in this grain. From 2 to 9 per cent of oil can be extracted from it, according to the variety employed.

In the distilleries of the Western States, where corn-brandy is made, the oil is extracted and sold as a product of the manufacture. A hundred bushels of the large southern or western corn, in which the horny part is not very large, yields fifteen or sixteen gallons of oil, which is at the rate of about 2½ per cent of the weight of the grain as it comes to market. Previous to distillation, the Indian corn is fermented with malt, and, during the fermentation, the oil rises to the surface and is skimmed off. It is a bright pale yellow agreeable-smelling oil, and sells for about a dollar a gallon. It is used for burning in lamps in western New York, in Ohio, in Michigan, and upon Lake Superior.

The popping of corn is owing to the presence of this oil, but only those varieties will pop in which the horny part is large, and surrounds completely the internal starchy part. The varieties called pop-corn and rice-corn possess this property in the highest degree. When they are heated in a close iron vessel—like a coffee-roaster—to about 600° Fahr., the oil contained in the horny part expands—perhaps in part decomposes—tears asunder every little cell in which it is contained, bursts the epidermis of the top or side of the seed with a slight report, like that of a popgun—and forces back, in fact turns outside in, the swollen and now white and spongy mass into which the horny part is changed. In this state the corn is soft and agreeable to eat, more easy of digestion, and is largely consumed. The increase of bulk by this heating process is so great that one barrel of pop-corn will produce sixteen, and of rice-corn, which is a small seed, thirty-two barrels of popp*ed* corn.

It will occur to the reader, from what I have said of the internal structure of Indian corn, that the flour which is obtained from the several varieties will be more or less yellow, according as the proportion of the coloured horny part is greater or less. Hence the white Tuscarora corn, which contains scarcely any horny matter, gives a whiter flour than almost any other variety. This is the variety, therefore, which is principally made use of for the manufacture of starch, and for the adulteration of wheaten flour.

As the direct fattening property of seeds is believed to be intimately connected with the quantity of oil they contain, I may mention in this place an interesting physiological fact, communicated to me by Dr Charles Jackson of Boston, which is susceptible of an important practical application. The horny part of the corn, he informs me, is not digested by the horse, though it is readily digested by the pig and by fowls. The economical value of a food, therefore, as I have elsewhere explained, cannot be judged of solely from its chemical composition.*

On our arrival at Rome, the neighbourhood of the station was crowded with people who were waiting to get a peep of Mr Clay, whom we had picked up at Utica on his way to Syracuse. Our train had now swelled to

* See the Author's *Lectures on Agricultural Chemistry and Geology*, 2d edition, p. 1045.

a line of fifteen large cars, containing each from forty to sixty passengers, so that we proceeded very slowly, and had become somewhat impatient of delay. A few cheers from the assembled crowd, rather faint compared with those which afterwards burst forth at some of the succeeding stations, were the only demonstrations of affection towards Mr Clay which I observed among the western Romans.

On leaving Rome, we forsook the valley of the Mohawk, and, in a south-westerly direction, crossed the Clinton group of green, sandy, ferruginous and calcareous shales, which, from their softness, have been much washed away, when the old sea-currents swept over them, and now form a flat, uninteresting, somewhat swampy country, stretching in a narrow zone along the whole of western New York, as far as the Falls of Niagara, and thence into Upper Canada. The largest and deepest depression in this belt of country is occupied by Lake Oneida, which we passed a few miles to our right, and by the marshes of the town(ship) of Cicero, which extend farther towards the south.

Nine miles from Rome we passed Verona, another memento of Italy; and a few miles farther, Oneida station, where we rapidly crossed a narrow belt of the Niagara group, the first of the upper Silurian system, and entered upon the Onondaga salt group, the most economically and agriculturally valuable of all the rocks of western New York. The natural softness of these groups of rocks in this locality, and the level character of the whole country, may be judged of from the fact that the Erie canal runs through it for sixty miles without a single lock.

A single glance at this country, from the time we left Verona, showed into how different a region we had come since we had left the Mohawk Valley. A flat forest country of mixed wood, with few clearings, resting chiefly on beds of gravel, accompanied us for some miles from the river; but before we reached Oneida we were upon soft shales, which crumbled into a tenacious soil. Rich red soils and red marly rocks succeeded; the country became cleared, was cultivated to the hill-tops, and appeared in a state of nature only where the flat and swampy surface and the clayey character of the soils rendered previous drainage necessary to successful cultivation.

From Oneida to Syracuse is twenty-four miles. During the latter half of it, and especially in the town(ship) of Manlius, we passed through much low, flat, sandy soil, still under forest. In some of the hollows, thick layers of peat rested on, or alternated with, the white drift sand; but where knolls and gravel hills occurred, Indian corn grew well; while the slopes of red land which skirted the valley on our left were covered with Indian corn, or with rich green herbage, to their very summits. We were now entering upon the wheat region, the old Genesee country, the ancient inheritance of the Six Nations. We reached Syracuse at half-past three, having come from Albany, 178 miles, with a constantly increasing train. Great crowds thronged the station and streets, and the city was a scene of much bustle and excitement.

CHAPTER VI.

The city of Syracuse.— Its rapid growth.—Popularity of Mr Clay.—His reputed chance, and that of Mr Webster, of the Presidentship.—Show of the New York State Agricultural Society.—Agricultural implements. —What they teach.—Forks, corn-shellers, and reaping-machines.—Extensive use of the latter in the North-Western States.—Argument against thorough drainage in western New York.—Want of local attachment.—Law against long leases in the state of New York.—Prevalence of the Devon breed of stock in New England, and of the Teeswater in the Western States.—Merino sheep.—Fast-trotting horses.—Over-lightness of the horses for heavy farm-labour.—General impression as to the condition of New York agriculture.—Crowds who visited the Show-yard.—Fruit Show.—Fruit region of Western New York.—Comparative mildness of its climate.—Rapid extension of apple-orchards.—Profits of apple-growing.—Quantity of fruit exported.—Pomological Convention.—Varieties of apples in the United States and in Normandy.—Mode of causing apples to produce a crop every year.—Influence of crops of rye on the apple-tree.—Influence of geological structure on the flavour of the apple, and of the cider made from it.—*Gout de terrain.*—Mr Geddes's farm.—Rich soils of the Onondaga salt group of rocks.—Soil of the crumbling green shale.—Rotation followed upon it.—Gradual exhausting effects of this rotation.—Average produce of the whole State of New York and of the richest western county.—Competition of the Western States.—Profits of farming in New York.—Property confers no political privilege.—Indian-corn hay.—Experiments with plaster or gypsum upon Indian corn and potatoes.—Gypsum acts best on calcareous soils and in droughty seasons.—Wages of farm-servants.—Escarpment of the Helderberg limestone.—Onondaga salt group.—Rich belt of land formed by it.—Section of the wheat region of western New York.—Beautiful relation of the soils to the nature of the rocks of which this undulating plain consists.—Soils of the Medina sandstone, of the Clinton, Niagara, Onondaga, Helderberg, Hamilton, Genesee, and Portage groups.—Influence of overlying drift.—Salt springs.—

Connection of gypsum and common salt.—Strength of the Brine Springs at Syracuse.—Quantity of water pumped up and of salt manufactured.—Alleged consumption of salt in the United States.—Individual consumption in Great Britain.—State revenue from the Salt Springs.—Methods of extracting the salt at Syracuse.—Probable presence of bromine in the Syracuse brines.

Syracuse, 12*th Sept.*—The village of Syracuse, erected into a city in 1848, occupies an important local position, and is a remarkable place in many respects. It is situated at the junction of the Oswego canal (of thirty-eight miles in length) with the main trunk of the Erie canal, and is thus on the great lines of communication between Canada and western New York on the one hand, and between the Atlantic and the Western lakes and States on the other. It is also close to the site of the salt springs, and is the seat of the extensive salt manufacture by which western New York, the Canadas, and the Western States are principally supplied with this necessary article of consumption. It stands also in a fertile district, and in a comparatively genial climate, where grapes ripen in the open air, and can even be left uncovered all the year through. Thirty years ago, Syracuse was the name of a few houses in the wilderness, now it is a city of 16,000 inhabitants, taxing themselves for local purposes during the present year (1849) to the extent of 35,000 dollars. It has the large hotels common to towns in this country, numerous churches, the spires of several of which are now in process of erection, and many skeleton streets, which, if its prosperity continue, will soon be built up. The last ten years have added greatly to its size; and, so late as six years ago, the wilderness still surrounded the residence of the mayor—to whose hospitality I was indebted during my stay at Syracuse—where his garden now extends, and plum and peach trees and vines are in full and luxuriant bearing.

All was excitement in the town, in consequence of the

arrival of Mr Clay; and complaints were not unjustly made by those engaged in preparing for the Agricultural Show that his presence, as a politician, seriously interfered with the objects for which so many thousands had assembled at Syracuse. To me, as a stranger, it was interesting to observe how very popular Mr Clay appeared to be, and how, in their liking for the individual, so many seemed willing to forget the shade of politics he represented. One of his political opponents observed to me, that, since the days of Washington, probably no man had so generally carried with him the hearts of the whole people of the United States. He is also a man of great ability, and has played a large part in the public affairs of his day; and it therefore appears remarkable to those who are unaware of the small matters by which great questions are decided, that he has never attained the dignity of President of the United States. He is now advancing in years, but is still fresh in intellect and full of energy, as is shown by his recent action upon the slavery question; and his friends are not without hope of seeing him still attain to that distinguished office before his intellectual strength is gone.

Of the able and ambitious men who now aspire to the office of President, there are many who think Mr Clay's chance all the better that the State of Kentucky has not hitherto had the honour of giving a President to the United States. It is a striking circumstance, that, of the twelve Presidents whom the Union has had, no less than five have been Virginia men; while Massachusetts and Tennessee have each sent two, and New York, Ohio, and Louisiana each one, to the Presidential chair. The principle of an equal division of office, or turn and turn about, is a very popular one in some of the States; and it has been gravely urged to me, by a person of much intelligence, in reference to Mr Daniel Webster, that the circumstance of his being from Massachusetts—a State which has already given two Presidents to the Union—is certain for ever to exclude him from that high position. But it is one of the benefits and boasts of a large federal republic, that a wider field exists from which to select great men to manage great affairs, and that it opens a wider field of ambition to the noble minds which may spring up in every part of the Union. Even in republics, however, the most excellent theory cannot be made to coexist with perfectibility in practice; and the alleged wider field for great talents becomes null, if the great offices are to be equally divided among the several States as their turn comes round.

13*th Sept.*—This morning I visited the show-yard, along with my friend Professor Norton of Yale College, who had thus far accompanied me in my tour through his native country. The show was held in a large inclosed area, quite as spacious as those usually devoted to this purpose by the Royal Agricultural Society of England. The two main divisions of implements and stock occupied the chief place, as with us; farm and dairy produce, however, and fruits, receive much attention from the New York State Society, and had an appropriate place assigned to them under the tents and sheds which were scattered over the grounds.

The general character of the implements was economy in construction and in price, and the exhibition was large and interesting. I know of no more instructive lesson, in regard to the practical condition of the husbandry of a country, than that which a man gets in surveying a collection of implements—actually in use, or coming into use—such as these exhibitions supply. Our English chaff-cutters and food-crushers, and drill and thrashing and tile machines, and cultivators and subsoil-ploughs and clod-crushers, tell more of what is going on in the country than months of travelling would make known to the most active agricultural inquirer. It is not so much the construction and cost and usefulness of a machine, as the number and variety of each seen on the ground, which shows what implements are in most request, and in what direction the practice of the farmer is progressing.

Ploughs, hay-rakes, forks, scythes, and cooking-stoves, were very abundant, and many of them well and beautifully made. American ploughs are now exported in considerable numbers. At a subsequent period, a dealer in Boston informed me he had this season sold a hundred of one of the varieties made in Massachusetts to a single individual for sale in London. The potato grips and forks, of various kinds, cut out of sheet-steel, were very elastic, light, strong, and cheap. They seemed to leave nothing in these articles to be desired. The cradle-scythes were also excellent: an active man was said to be able to cut four to six acres of wheat a-day with them. That, of course, would depend something upon the quantity of straw upon the ground.

Among the more novel instruments to me were the corn-shellers and crushers. The former were very pretty implements, and the larger kinds were said to be capable of shelling two hundred bushels of Indian corn an hour.

Of reaping-machines there were several varieties on the ground, and several are actually in use in the Western States. Hussey's, which I saw on the ground, was said to cut twenty-five acres of wheat a-day. My friend, Mr Stevens, who went round the yard with me, assured me he had seen one of them cut sixteen acres. M'Cormick's machine, I suppose, must be a good one, from the information here given me that as many as fifteen hundred of them have been made at Chicago, in Illinois, this last year, and sold for cutting wheat on the prairies of the North-Western States. Of course, it is only on flat lands that they can be advantageously employed. But where labour is scarce, and unwooded prairie plenty, the owner of a reaping and a threshing machine may cultivate as much land as he can scratch with the plough and sprinkle with seed.

A great breadth of this western New York is wet, flat, and marshy; but drainage is yet unknown. A single sample of pipe-tiles was exhibited—as a curiosity, I suppose. But the Society is now alive to the importance of drainage, not only for the purpose of drying flat and obviously wet and marshy land, but of rendering more productive such heavy soils as with us are so generally improved by thorough-drainage; and already several tile-machines have been imported from England under their auspices, and premiums offered for experiments in thorough-draining.

An objection to drainage is made in this country which, though sometimes urged with us, is by no means of such force in England as it is in America. The cost of this improvement, even at the cheapest rate—say four pounds or twenty dollars an acre—is equal to a large proportion of the present price of the best land in this rich district of western New York. From fifty to sixty dollars an acre is the highest price which farms bring here; and if twenty-five dollars an acre were expended upon any of it, the price in the market would not rise in proportion. Or if forty-dollar land should actually be improved one-fourth by thorough-drainage, it would still, it is said, not be more valuable than that which now sells at fifty dollars; so that the improver would be a loser to the extent of fifteen dollars an acre.

This argument will appear to have greater force when it is understood that there is as yet in New England and New York scarcely any such thing as local attachment —the love of a place, because it is a man's own—because he has hewed it out of the wilderness, and made it what it is; or because his father did so, and he and his family have been born and brought up, and spent their happy youthful days upon it. Speaking generally, every farm from Eastport in Maine, to Buffalo on Lake Erie, is for sale. The owner has already fixed a price in his mind for which he would be willing, and even hopes to sell, believing that, with the same money, he could do better for himself and his family by going still farther west. Thus, to lay out money in improvements is actually to bury what he does not hope to be able to get out of his land again, when the opportunity for selling presents itself.

With us the mode of looking at improvement questions is different. A proprietor says, this is my own residence; I will make it as comfortable as I can—as valuable as I can—improve it as much as my means will allow—make it equal, if possible, to the best land of my neighbours. If I should spend a good deal of money upon it, I am depositing it in a sure bank, if I go prudently and skilfully to work; and my children, at all events, will reap the benefit.

Or, if he is a tenant holding the land on a lease for so many years certain, he balances the cost of drainage against the increase of produce, and the diminution of expense in working his land, taken together, during the term for which the holding is secured to him; and he drains, or the contrary, as this balance proves encouraging or otherwise. The owner in New York, who also farms his land, should consider the question of improvement as the English or Scottish leaseholder does, were it not that he is in reality a less fixed being than the tenant-farmers are in our island. I do not know whether we ought to consider that the moral force which originally projected them or their fathers from Europe is still partially unexpended, and thus tends, at any unwary moment, to bear them still farther off; or whether, as some think, the climate of North America creates in the Anglo-Saxon race a nervous restlessness which continually incites to change;—but certainly "forwards," in a physical at least as much as in a moral sense, is the characteristic of our respected Yankee cousins.

I am reminded, by my allusion to the leasehold tenure, of a singular provision in the New York State Constitution in reference to leases. The 14th section of the 1st article says:—"No lease or grant of agricultural land, for a longer period than twelve years, hereafter made, in which shall be reserved any rent or service of any kind, shall be valid." This section owes its origin to the existence in this State of certain old leases, by which a small quit-rent was reserved, and which leases the present leaseholders have been anxious to convert into freeholds, not by buying out the manorial rights, but by refusing any longer to pay the quit-rent—as if it were really a hardship to fulfil honestly the terms on which their lands are held. The clamour and excitement to which this struggle gave rise was, no doubt, the main cause of the insertion of this restriction into the Constitution as to the length of future leases. The time may come when, if not altered, this law will operate as a bar to agricultural improvement. In Scotland, a shorter lease than nineteen years is not considered to give sufficient security or inducement to the tenant to make those expensive improvements which his landlord does not find it expedient to make with his own capital. The time is no doubt approaching, in this commercial State, when the letting of farms, on money or grain rents, will be both more general and more popular than it is at present; and then, to secure good farmers, capable of expending the necessary capital, a change in this article of the Constitution may facilitate that advancement in agriculture which, as far as circumstances admit, so many parties are already eagerly striving to make.

In the New England States and in New York the Devon blood prevails. Most of the stock are *grades*, as they are called, or crosses of the pure Devon bull with the older stock of the country, which is originally of mixed English and Dutch of various kinds. The cows exhibited were nearly all Devons, and there was a beautiful Devon bull in the yard which had been bred in Canada. In the Western and South-Western States the short-horn blood predominates, and of this blood there were some good specimens exhibited.

The Merino sheep are great favourites, and in the Rembouillet stock the carcass has been much improved. If they have dry lying, they stand the winter well in open sheds.

Of horses there was a large show. Of that fast-trotting horse which is so much fancied in this country there were many exhibited. It is in the exigencies of a new country, where few horses could be kept by the small farmers, and the necessity for having them of a kind which could both work in the field and go fast to market, that we find the origin of that general lightness of body which distinguishes the Canadian and other North American horses. They are, in reality, too light for heavy farm-work; and when the period arrives for deep-ploughing, and more extensive cultivation of heavy land, a heavier and stronger stock of horses, still preserving a quick step, will gradually take the place both of the oxen, which, in many of the States, are now extensively employed, and of the limber-horses, with which they are sometimes yoked in the same team.

On the whole, the opinion I formed of the actual condition of New York agriculture, from the show of implements and stock on this occasion, was a very favourable one. That the practical agriculture of the State is far behind that of the best parts of England and Scotland no one can deny; but that there are good farmers in the State, that progress is making generally, that a desire for progress is very widely diffusing itself, and that there are men of skill and energy at work in exciting and directing that progress, the occurrences of this meeting sufficiently testified.

The premiums offered by the Society on this occasion amounted to the sum of £1600, and the number of persons who visited the show-yard was immense. Thirty thousand entrance tickets (sixpence each) were sold on the first day of the show, besides three thousand members' tickets, (one dollar each,) which admitted the member and his family.

At three P. M. I delivered my address, in a large open tent, to an audience of several thousand people, by whom it was warmly and kindly received.

The fruit-show was not so fine as was anticipated, owing to the season having proved an unfavourable one for fruit in the Northern States generally. By the residents of western New York this is regarded as the finest fruit-country in the world. The mollifying influence of Lake Ontario—which has an area of 6300 square miles, an average depth of 500 feet, and never freezes as Lake Erie does—extends, more or less, over the whole level or slightly undulating region occupied by the lower portion of the upper Silurian rocks, on which the rich soils of this part of the State rest, and from which they are generally formed. From Oswego, near the east end of Lake Ontario, to Niagara, on the Canadian borders, this region forms a belt about 40 miles wide by 150 miles long, and over it the early frosts of autumn, which are so injurious to fruit-trees, are comparatively unfelt. It is on the eastern part of the Lake Ontario shore, towards Oswego, however, that the grape and peach ripen the most surely, and produce the finest fruit.

The old apple-country of the United States—the home of the Newtown pippin, the Spitzemberg, and other highly prized varieties — is on the Atlantic border, between Massachusetts Bay and the Delaware. But western New York and northern Ohio have now entered into earnest competition with these old orchard countries. Their rich soils produce larger and more beautiful fruit, but inferior, it is said, in that high flavour which distinguishes the Atlantic apples. This inferiority, however, is not conceded by the western cultivators, among whom orchard-planting is rapidly extending, and who estimate the average profit of fruit cultivation at 100 to 150 dollars an acre, (£20 to £30.)

In Oneida County, at the eastern extremity of Lake Oneida, part of four townships shipped on the canal, in 1848, as many as 18,000 barrels, at from 62½ to 100 cents per barrel. This is a low price for good apples. But in New York the best apples sell for three or four, and in London for nine dollars a barrel. In Wayne County, about the middle of this belt of land, the merchants of Palmyra, a shipping village on the Erie Canal, sent off 50,000 barrels of green or fresh, and 10,000 of dried apples in the same year, and along with them 1000 bushels of dried peaches.

These facts show that the fruit-culture is becoming one of considerable importance, and the number of persons now interested in it has caused the formation of what is called the Pomological Convention, which held its meeting at Syracuse on the occasion of the State agricultural fair. Everything connected with the culture of the apple, and, I believe, of the other more common and more important fruits, such as the pear, the peach, and the grape, form subjects of inquiry, consideration, and discussion; and from the proceedings of this Society good can scarcely fail to arise to this growing branch of practical agriculture.

Nearly 200 recognised varieties of apples are already cultivated in the States, of which 186 are described by Mr Downing.* One important object

* *Fruits and Fruit-trees of America.* New York, 1849.

which the American Pomological Convention may usefully keep in view is the purification of the nomenclature of fruits. In Normandy, where the apple-culture for the manufacture of cider is an important source of revenue upon every farm, there are upwards of 5000 differently named varieties of the acid or bitter apple, which yield the cider. These have all been collected and examined by Professor Girardin of Rouen, grafted on stocks, grown, figured, and analysed, and he informed me that the same apple was sometimes known by as many as eighteen different names in different parts of that country. I was struck, during a tour in Normandy two years ago, with the little skill and the antique and rude tools which appear to be there devoted to a branch of industry which is of much economical value to the farmers of the province. There are no hedgerows upon the farms, but the divisions of the fields are marked out by rows of apple-trees; and the crop of fruit, which there, as in so many other countries, is good only every second year, is expected to pay the whole rent of the land. In behalf of an industry of so much consequence to the rural population of a large province in France, to whole counties in England, and to large portions of other countries of Europe, and which is likely to be extensively prosecuted in America, it is time to ask whether the sciences of botany, meteorology, chemistry, and physiology cannot now be made to render them more direct economical aid than they have hitherto done.

In the United States only the finest apples are sent to market—the waste or refuse are generally made into cider. But those varieties which are best for the table are unfit alone to make a palatable cider. The culture, growth, and selection of cider apples, the proper admixture of varieties in the crushing-mill, &c., is a branch of special husbandry requiring special knowledge, the acquisition and diffusion of which may be greatly promoted by a judiciously conducted association of growers.

In the United States also, as elsewhere, the trees naturally yield a heavy crop only every second year. But Mr Pell, the owner of one of the finest orchards in America—that of Pelham farm, at Esopus, on the river Hudson—to whom I was subsequently indebted for much kind attention at New York, has recently been making experiments with the view of ascertaining whether, by proper manuring applications to the roots, an annual crop might not be secured from his valuable Newtown pippin trees, of which he has two thousand in full bearing. His experiments, he informed me, had been perfectly successful, only he had begun to think, or apprehend, that the life of his trees might be shortened by this course, and that he might have to replace them so many years sooner. But should this prove the result, it might still be profitable, as it is with the peach orchards of New Jersey, to have a succession of new trees coming up to replace the old—and experiments on the subject are deserving of encouragement.

An interesting observation made by Mr Pell in regard to the influence of crops of rye upon an apple-orchard is deserving of a place among important physiological facts as yet incapable of explanation. He cultivates his orchard grounds as if there were no trees upon them, and raises grain of every kind except rye—which crop he finds so injurious that he believes three successive crops of it would destroy any orchard which is less than twenty years old. We in Europe, who think it bad and exhausting husbandry to take three successive corn crops of any kind from the same land without the addition of manure, might be inclined to attribute the ruin of an orchard, under these circumstances, to bad husbandry only, were it not that similar successive crops of other kinds of grain do not produce a like effect.

How interesting it would be to follow out the practical physiological experiments which this observation of Mr Pell suggests!

It is a curious illustration of the connection of geology even with this branch of rural industry, that the nature of the rock over which the apples grow affects the flavour of the cider which is made from the fruit. The cider of the chalk soils in Normandy differs in flavour from that of sandy, and both from that of clay soils—the variety of fruit and the management being the same. Hence the *gout-de-terrain* spoken of by French connoisseurs is a correct expression for this recognisable difference. Among the varied geological deposits of western New York, similar differences must likewise be observed both in the fruit and in the cider made from it, which will give peculiar characters and recommendations to the productions of the several districts.

14*th Sept.*—This morning early, the Honourable Mr Geddes, one of the State senators for Onondaga County, drove me out to breakfast at his farm of Fairmount, a few miles from Syracuse. He is the owner of 300 acres of the best land on the rocks of the Onondaga salt-group, and, like nearly all the owners in this country, lives upon and farms it himself. The soil is a light-coloured calcareous clay, which crumbles readily, and never bakes. It is generally shallow, and rests on a green shaly rock, which readily crumbles in the air, and by exposure becomes paler in colour, forming the light-coloured soil of which the farm consists.

This district is more like a part of Old England than of a newly cleared country. Of Mr Geddes's 300 acres 270 are in arable culture, and comfortable houses and other buildings are seen upon most of the farms. The land is generally divided into farms of one or three hundred acres; and these, with the buildings upon them, usually sell at from fifty to sixty dollars an acre. At this price

Mr Geddes considers it the cheapest land in the States—for those of course who have the money to buy it. By men whose capital is in their bodily strength and industrious habits, the wilderness land of more western districts is alone attainable.

The land is of a very useful kind, producing all sorts of grain crops well, though not of equal quality. Thus the produce per acre was stated by Mr Geddes to be, of—

Wheat,	18 to 35	bushels	of 60 pounds.
Barley,	20 ... 55	...	of 48 ...
Oats,	40 ... 100	...	of 32 ...
Indian corn,	50 ... 80	...	of 56 to 60 pounds.
Potatoes,	100 ... 300	bushels.	

It is least adapted, he said, to the growth of potatoes; and turnips are as yet but little grown, as the raising of fat stock is not much attended to. An average weight of 32 lb. does not indicate a soil or climate well suited to the oat crop.

This land has been in many places ploughed for fifty years without receiving any manure. I walked over two large fields which have never been manured for the fifty years which have elapsed since the present owner's father cleared them; and he thinks the land still as good as ever it was. He reaps from it 50 to 60 bushels of corn; and, last year, (1848,) 30 bushels an acre of wheat. The soil consists, for the most part, of crumbling fragments of the green shale. When the older land appears to become exhausted, the plough is put in a little deeper, so as to bring up a little of the crumbling rock, (green shale,) when it is said to produce wheat as before.

The rotation is Indian corn after lea, with manure—if any is applied—then oats, followed by barley or pease, and finally, winter wheat, with seeds in spring. It is kept in grass two years; in one of which two crops of hay are cut, and in the other it is pastured, as 250 sheep and 16 milk-cows are kept. If the land be foul, it is now summer fallowed, and sown with wheat, followed by seeds as before, after which Indian corn comes again. If it is not foul, the rotation commences with Indian corn after the first two years' grass.

On land like this extraordinary green shale land, such severe—what we should call scourging—treatment may be continued a great many years with apparent impunity; although it tells very soon on land of inferior quality. But even on such land it tells at last. Hence it is that this celebrated wheat-region, *as a whole*, is gradually approaching the exhausted condition to which the more easterly wheat-growing, naturally poorer districts, had earlier arrived. They are ceasing, in many places, to be remunerative in the culture of this crop with the present system of farming, are becoming unable to compete with the cheap wheat-growing virgin soils of the West, and are therefore in such places—as I was informed on the spot—gradually being laid down to grass, or turned to other more promising agricultural uses.

The average produce per acre of the whole State of New York, as published by the State Agricultural Society,* is, for—

Wheat,	14 bushels.	Oats,	26 bushels.
Barley,	16 ...	Indian corn,	25 ...

Potatoes, 90 bushels, or about $1\frac{1}{2}$ tons an acre.

The averages for Monroe County, in the middle of this western district, are the highest; and they are as follow:—

Wheat,	$19\frac{1}{2}$ bushels.	Indian corn,	30 bushels.
Barley,	19 ...	Potatoes,	110 ...

Oats, . . . 32 bushels.

For a highly lauded, fertile, wheat-growing district,

* *Transactions of the State Agricultural Society*, for 1845.

the pride of the State of New York, the happy home to which the longing eyes of British and Irish agriculturists have long been directed, these are but low averages. Either the land is not so good as it has been called, or it is and has been badly treated. The treatment has certainly been bad; but, as surely, large portions of the land are naturally very good, and may still be made very productive. But these farmers of western New York are exposed to the competition of their more western brethren—a still nearer competition than that under which our British farmers are suffering; and it is doubtful whether, with all their naturally rich soils and low-priced lands, they can, on the longer cultivated and more exhausted tracts, successfully struggle with them in the raising of wheat. If they can, it must be as with us at home, by the application of more skill and prudence than has hitherto been applied to rural affairs.*

According to Mr Geddes, farming is so profitable in this part of the State, that a person can buy 100 or 150 acres of land, can pay a part of the price in ready money, give a mortgage for the remainder at 7 per cent, can pay this interest, and gradually clear off the debt. This, he said, was a very common practice; and that hence many of the farmers had mortgages on their land. This may be true of Mr Geddes's neighbourhood; but all I have heard from others of the general condition of the existing rural population north and east of the Delaware, and of Lake Erie, is in confirmation of the opinion that money is not to be made by farming, at least in these parts of the States. I do not speak at

* "What, I ask, is to meet this competition of the west, but greater skill and greater care in the mode of agriculture?" These words, addressed at the close of 1850 to the Oswego Agricultural Society by their president, are a re-echo of what has been so often said among us at home.

present of the more western States, which I have not myself visited. A comfortable livelihood and adequate support for a family may be obtained by ordinary industry, but money is accumulated with difficulty; and this is the test of prosperity which all classes apply to their pursuits. Hence those who wish to add to their capital more readily, or more speedily, betake themselves to traffic, or to some other more promising employment. Hence, also, one reason why so many farms are in the market. The price of land rises as a district becomes settled; so that, when a man's sons grow up, and are ready for farms of their own, he is unable to provide for them by purchasing land in his own neighbourhood; but by selling his own clearing at the increased value it has acquired, he can proceed further west, and, with the price he receives, provide farms in the wilderness for them all.

The renting or hiring of land, especially for a money rent, is not more popular in this district than elsewhere, even on a lease. Farmers do not like to be tenants; and when land falls into the hands of mortgagees, and must be let, it is usually let on shares, sometimes on halves, as it is called; sometimes at two-thirds of the produce, as the agreement may be, which is special for almost every case.

The usual size of farms is from 100 to 150 acres. Some farms are as large as 1000 acres; and a family of Monroes was mentioned to me in this neighbourhood who have farms of this size which they cultivate and manage themselves. A large landholder, unless he farms all his land himself, is looked upon with dislike as an aristocrat. Property, according to the constitution, confers no political rights, except upon coloured men;* and there is a jealousy lest a man, by

* A free coloured man, in the State of New York, must possess a freehold of 250 dollars before he is allowed to vote as a citizen.

having others under his influence as a landlord, should acquire a proportionate increase of individual political power. Hence, if a man does not wish to cultivate any part of his land, or cannot do so, he sells it, and invests the proceeds at 7 per cent, which is readily obtained. He is, then, whatever his wealth may be, no longer exposed to odium or dislike.

I went into a fine field of Indian corn on Mr Geddes's farm, upon which his men were at work reaping. The stalks are grasped in the left hand, and cut down near the ground by means of a heavy sharp hook, resembling a bill-hook. They are then tied together in sheaves and set up to dry; after which the sheaves are carried to the barn, and the head of corn separated by the hand. The corn-sheller then quickly separates the grain from the cob or internal head-stalk.

The very wasteful practice exists in many parts of North America of reaping the Indian corn very high, so as to leave as much as two feet of the stalks in the ground; and in others, of altogether rejecting the stalks when reaped, using them neither for food nor in making manure. But cattle eat the corn-stalks very willingly: they are said by some to be equal to the best hay in feeding, and to produce more milk than hay does. Thus, in addition to the grain, an acre of land yields three tons of fodder, which saves much other food; and when cut by means of a chaff-cutter, may be made the means of procuring a large supply of manure. Wherever the exhausting system has nearly done its work, and the value of manure has begun at last to be appreciated, attention will, by degrees, as is now in some measure the case in New England and New York, be drawn to the great value of the hitherto neglected corn-stalk. Of course, the degree of ripeness to which the corn is allowed to attain, the variety sown, the soil on which it grows, the fierceness of the sun beneath which it ripens, and other circumstances, will affect its value as a nutritive fodder, and will, in different districts, modify the way in which it can be most usefully or most profitably employed.

Much has been said at different times about the introduction of Indian corn as a field-culture into England—an object which, I fear, our feeble summer heats, cloudy days, and early frosts, will prevent us from ever extensively attaining; but as a green food to be cut in its unripe state, and given green or dried for winter, it might be introduced with a chance of profitable success.

Plaster or gypsum is extensively used in this neighbourhood, being almost the only manuring which a large portion of the land receives. It is obtained abundantly among the beds of the Onondaga salt-group, and is applied in the unburned state. It is crushed in mills, where it is sold in the state of powder at 3d. a bushel, or a dollar and a half a ton of 25 bushels.

The maize is plastered either once broadcast, at the rate of 3 bushels an acre, or twice with the hand, upon each hill after each hoeing, at the rate of 1 bushel an acre. I saw four rows in the fine field of Indian corn I walked through which had not been plastered, while all the rest had—once only at the rate of a bushel an acre; and the difference in favour of the plastered part was very striking to the eye. Oats are also much benefited by plaster, especially in a dry season like this; and it brings away clover, and makes it very tall. It is likewise believed to improve the potatoes which are planted without manure. I caused a number of plants of the potatoes to which gypsum had and had not been applied to be dug up, and certainly the number and size of the potatoes found at the roots, as well as the height of the stems, were greatly in favour of the plastered part. It was applied before hoeing, and then drawn up around them. It is usual to put it in with the sets; but it was put around the young plants this year, "only because," said Mr Geddes, "the drought was such that I saw if something was not done I should have no crop at all." An English farmer would hardly believe he had done anything towards saving a crop of potatoes if he had only sprinkled a bushel of gypsum over an acre of the land in which his potatoes were growing.

This beneficial action of gypsum, notwithstanding all that has been written upon the subject, is still very wonderful, and not the less so that in so many places, and in so many circumstances, it fails to produce any sensible effect. Mr Ruffin, states that in the Carolinas it is found to produce the best effects upon land which has been already limed; and here, where its beneficial influence is so manifest, the land is naturally rich in lime. I have caused an analysis of a portion of the green shale from which the soil on Mr Geddes's farm is formed to be subsequently made, and have found it to contain as much as 23 per cent of carbonate of lime, and 13 per cent of carbonate of magnesia.* It may be, therefore, that while this marly and magnesian character of its soil certainly makes the district more favourable to the growth of wheat, that it has some influence also in disposing it to be beneficially acted upon by gypsum. This substance does not appear to require rain to aid its effects, since it is applied especially in

* The composition of this slaty rock from Mr Geddes's farm at Fairmount, near Syracuse, was as follows:—

Water of combination,	3.02
Alumina,	9.85
Oxide of iron,	2.87
Carbonate of lime,	23.22
Carbonate of magnesia,	13.81
Carbonate of iron,	1.45
Insoluble siliceous matter or silicates,	46.13
	100.35

The presence of phosphoric acid in this sample was not appreciable.

droughty seasons. A calcareous soil and a hot sun may possibly, therefore, be instrumental towards its success.

This view is further supported by the prevalent opinion among farmers "that the great use of plaster is *to draw water from the air;*" which means, as I take it, that its action is more apparent in dry than in wet seasons.

A very extreme view of its influence upon the weather was entertained by some of the old Dutch farmers in the United States—one of whom, according to Judge Peters, objected to the use of it because "*it attracted thunder.*"

Mr Geddes works his farm with four pairs of horses and seven men, on an average, all the year through. His head-man has 313 dollars a-year. Other men have ¾ dollar (3s. English) a-day, except in harvest, when they have 1¼ to 1½ dollars a-day. A good man, taken into the house, has 150 dollars (£31, 10s.) a-year and his board. They are hired by the month.

Behind Mr Geddes's farm, at a short distance towards the south, rises the escarpment of the Helderberg limestone, the outcrop of which, more or less distinct and elevated, runs east to the Hudson River, and west as far as Lake Erie, and forms the southern limit of the belt of low rich land of which this wheat-region consists. He drove me in an open carriage for some miles along this escarpment, and thus enabled me to obtain a general idea of the whole country, and gave me an opportunity of picturing to myself what this broad plain will ultimately become when arterial and thorough drainage have done their work, and the axe of the clearer has laid open the broad patches of wilderness which still stretch, with occasional wide breaks, on almost every side as far as the eye can reach.

The Onondaga salt-group, on which Mr Geddes's farm reposes, consists on the surface of green shales richly calcareous, sometimes impure shaly limestones, to which succeed similar shales containing deposits or rounded nests of gypsum. These rest on a porous limestone, beneath which again occur green and red shales, calcareous and crumbling like those above, and like them forming rich wheat-soils. It has an average thickness of about 1200 feet, and forms a belt of generally level and undulating land, running east from Syracuse nearly a hundred miles, and west as far as Buffalo, and again beyond the Niagara River, far into Upper Canada. The breadth of this belt at Syracuse is about ten or twelve miles, and it rapidly tapers off to nothing as we go east towards the Hudson River. But westward it first expands, in Seneca and Wayne counties, to a breadth of between twenty and thirty miles, and afterwards continues from sixteen to twenty miles wide for nearly one hundred miles, after which it contracts to about twelve miles, which is its width on the river Niagara. This formation alone, therefore, forms a large area of rich land. But the country to the north of it, as far as Lake Erie, is also underlaid by rocks which crumble readily, and yield soils of good quality, and generally rich in lime; while to the south the nature of the rocks, and the agency of those causes to which the spread of drift is owing, have both in part contributed to the production of good grain-growing land. Hence, in the so-called wheat-district of western New York, is included the broad belt of about forty miles in width—from the shores of Lake Ontario, southwards to and including a large portion of the Hamilton shales of the New York geologists. As this country presents a really interesting illustration of the influence of geological causes on agricultural capabilities, I subjoin a section across this wheat-district at its broadest part, where it includes the eastern portion of the county of Wayne, and a portion of the county of Seneca.

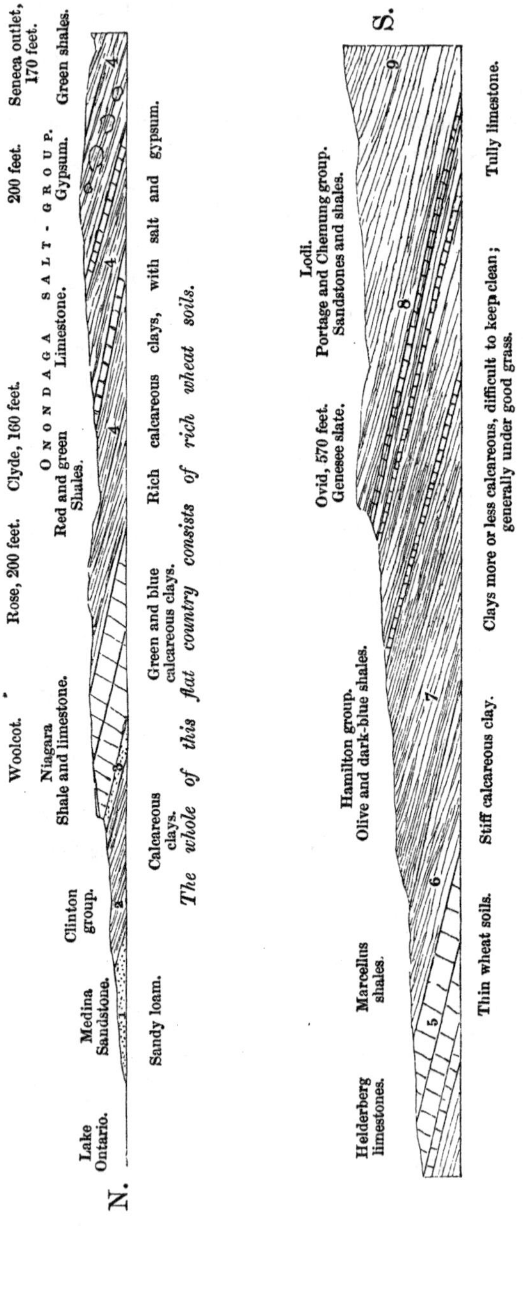

N.

Woolcot. Rose, 200 feet. Clyde, 160 feet. 200 feet. Seneca outlet, 170 feet.

Lake Ontario. Medina Sandstone. Clinton group. Niagara Shale and limestone. Red and green Shales. ONONDAGA SALT-GROUP. Limestone. Gypsum. Green shales.

1 2 3 4 4 4

Sandy loam. Calcareous clays. Green and blue calcareous clays. Rich calcareous clays, with salt and gypsum.

The whole of this flat country consists of rich wheat soils.

Ovid, 570 feet. Lodi.

Helderberg limestones. Marcellus shales. Hamilton group. Olive and dark-blue shales. Genesee slate. Portage and Chemung group. Sandstones and shales.

5 6 7 8 9

Thin wheat soils. Stiff calcareous clay. Clays more or less calcareous, difficult to keep clean; generally under good grass. Tully limestone.

S.

From the lake to the beginning of the Marcellus shale —No. 6 on this section, at a height of 216 feet, the level of Seneca Lake—is about thirty miles. The rise of the country, therefore, is very gradual; indeed, it is comparatively level after the first ascent over the Niagara limestone till we reach the Marcellus shale. Beyond this shale it rises more rapidly; and at Ovid there is a bold escarpment where the two bands of limestone occur. In other parts of the country there are also bolder escarpments along the out-crop of the Niagara and Helderberg limestones than the section shows, forming marked and successive steps as we advance into the interior.

From the lake to the Marcellus shale is the proper wheat-district, though in some places it extends to the Genesee slate, and may ultimately, under a better system of culture, be generally extended thus far. But the rapid rise of the land, by altering the climate, will always make this higher region less propitious to vegetation, supposing the soil as good as those of the Onondaga and more northern formations.

No. 1. The Medina sandstone consists of layers of brownish sandstone intermixed with shales. It presents an interesting illustration of a fact of much importance in agricultural geology, that the same formation at different parts of its extent may produce soils very different in their nature. Towards the east, the sandstone greatly predominates, forming sandy soils comparatively poor in agricultural value. But towards the west, the shaley or clay beds increase in number or thickness; so that sandy loams, and finally clay loams, and excellent wheat-soils, are produced from it.

No. 2. The Clinton group consists of green and blue shales and limestones, the admixture of the fragments of which forms an excellent wheat-soil.

No. 3. The Niagara formation consists of the soft shales below, and of the thick impure limestone beds above, which give their character to the Niagara Falls, and to the outline of the landscape. The shale alone forms stiff clays, which, from the sloping nature of the surface, are generally dry and susceptible of culture; while the limestone alone, where it is sufficiently crumbled, produces a surface adapted for wheat or Indian corn. But where, from the washing away of the Clinton beds, which are only 60 or 80 feet in thickness, the Niagara shales come in contact with, or are mixed with the debris of the Medina sandstone, soils are produced which are of "unequalled fertility"—illustrating another important principle, of which we have many examples in England, that, at the junction of beds of different kinds of rock, the soils are often much superior to those which are produced by the fragments of either rock alone.

No. 4. The Onondaga salt-group consists, as I have already said, of red and green shales below, succeeded by porous limestones, and these by beds of shale, including irregular but larger deposits of gypsum, the whole surmounted by other green shales or thin beds of impure light-coloured limestone, containing much magnesia. Calcareous matter abounds through the whole of this fertile formation—and generally the soils are rich, free, and easily worked.

No. 5. The Helderberg limestones and sandstones, where the surface is hard and rocky, are often covered with a thin soil of less value; but where the soil is deep, it is of excellent quality. Here, however, as a wheat-region, the natural quality of the surface begins to fall off.

No. 6. The Marcellus shale is thin, varying from a few feet to 60 or 80, so that its effect on the surface is seen chiefly by its improving the Helderberg series at the point of junction, and by forming occasional stripes and patches of stiff clay.

No. 7. The Hamilton group, when alone, forms stiff dark-coloured clays, which are less rich in calcareous matter than the Onondaga soils, and therefore less free and more difficult and expensive to work, but capable of producing excellent wheat. A large portion of the celebrated Genesee Valley rests upon this formation; but its natural soil is there covered or modified by the drifted fragments of the Niagara and other more northerly limestones. This group is of great thickness, and forms an extensive belt of country, the soils of which in some places are rich in lime, and are submitted to arable culture. They are everywhere difficult to keep clean, however, and are especially infected with the pigeon-weed (*Lithospermum.*) They are, for the most part, therefore—like our own stiff clays of the lias and other formations—left to perpetual grass, which they produce of excellent quality. Here, therefore, the grazing and dairy country of western New York commences.

No. 8. The black Genesee slate is too thin to form an important agricultural feature in the country. It crumbles more slowly than the Hamilton shale, but where it mixes with the thin limestones and calcareous shales beneath it, good soils are produced.

No. 9. The Portage and Chemung groups consist of alternations of shales, poor in lime below, with flagstones and massive sandstones. They extend to the borders of Pennsylvania, where they reach the height of 1000 feet above Lake Ontario. The district occupied by these groups presents a complete contrast to the wheat-region. When first cleared, it produces crops of wheat; but after the first crops—as is the case in many parts of New Brunswick, which rest upon similar rocks—wheat-becomes uncertain, and spring grain only can be sown. It is therefore poorer, less cleared and cultivated, and possesses a poorer race of cultivators. The farmers devote their attention chiefly to the rearing of stock, and to the dairy husbandry.

To teach a man the close relation of natural agricultural capabilities and early agricultural prosperity with the nature of the rocks of a country, it is only necessary to ascend from the valley of the Onondaga shales to the hills of the Portage sandstones.

But another interesting and instructive feature presents itself in this wheat-growing belt. The waters which in former times descended this region came from the north, and have drifted the materials of the more northerly over the edges of the more southernly rocks. Thus the materials of the Medina sandstone are made to overlap the Clinton and Niagara clays, and so to mingle with, lighten, and improve the soils formed from them. The Niagara shale, again, has overlapped the overlying Niagara limestone, mixed with it, and deepened its naturally thin and more slowly crumbling debris. So the various soft beds of the Onondaga salt-group have been intermingled, and a comparatively low, level, and undulating surface been given to the whole. The abundant materials derived from this easily crumbling group, again, have been spread over the Helderberg limestones, and the fragments of these over the Marcellus and Hamilton shales, adding to the calcareous matter of these latter rocks, and thus widening the belt of wheat-growing country beyond its natural limits.

This widening is especially visible along the north and south valleys, which penetrate far into the Hamilton and Portage groups of rocks, and into which the fragments of the Onondaga group were naturally carried by the rushing water. Up the valley of the Genesee River, and into the outlets of the Seneca and Cayuga lakes, this drift has penetrated farthest; and to its presence is due the peculiar agricultural excellence for which the soils of these localities are known, and which the rocks on which they rest could not alone have imparted to them.

The first freshness of nearly all these naturally fertile

lands, which have been long under cultivation, has been rubbed off them by ignorant and improvident culture; but skill, the child of greater knowledge, can restore them again to great productiveness. How far it will pay to improve even this country, for the growth of wheat—as our best English and Scotch farmers would like to improve it—while the still uncleared lands of the north-west are competitors in the wheat-market, I do not possess sufficient knowledge of local circumstances to enable me to decide.

With Mr Geddes I afterwards visited the salt-works for which this district is so celebrated, and to which the city of Syracuse is so much indebted for its rapid prosperity.

Salt-springs occur over the whole of this belt of country, from the Medina sandstone to the Portage group inclusive; and salt has been, or still is, more or less extensively manufactured from them. But they are most abundant, and yield most copious supplies of water, in the Onondaga salt-group, which derives its name from them. In this group they are generally richer also in saline matter, and yield a purer salt.

As in so many other localities, the gypsum appears to be connected in its mode of deposition with that of mineral salt. There occurs, indeed, in this locality—upon Mr Geddes's farm among other places—a very interesting proof of the close connection of these two mineral deposits. In the green shale-beds are found numerous pseudo-morphous masses, sometimes as much as six or eight inches in diameter, which appear to be casts of the interior of large hollow crystals of common salt, such as are formed, of a smaller size, on the surface of salt-pans, or of bodies of salt-water which are evaporated in the open air. Such hopper-shaped casts, as they are called, were very numerous in the locality in which I saw them, throughout the entire body of the shale, and seemed to show that when this shale was deposited in the state of a fine mud, it was impregnated with salt—probably was at the bottom of a salt lake, liable to be dried up or concentrated by evaporation, and that, when so concentrated, the salt crystallised among the mud in large crystals, as impure salts usually do, and impressed the shapes now seen upon the plastic mud. No solid salt, however, has ever been found in connection with these pseudo-morphic masses. It has no doubt been long dissolved out by rain and spring waters sinking through from above. But while the salt-springs which occur throw light on their formation, these crystals, on the other hand, explain in what way the upper Silurian strata of western New York have in so many places become impregnated with salt.

The wells or borings from which the brine is obtained are sunk in the marshy ground which borders the Onondaga Lake. This lake, which is about five miles long and one mile broad, rests on a bed of impervious marl, beneath which lies a thick deposit of drift, occupying a valley hollowed out of the shales of the salt-group. Into this drift the borings have been carried to a depth of nearly 350 feet; and the supply has generally been more copious, and the water more strongly impregnated with salt, the deeper the boring has penetrated.

If the quantity of salt contained in water which is saturated be represented by 100, that contained in the strongest springs near Syracuse is 76. A hundredweight of salt is obtained from about 700 gallons of sea-water; but of the water from these springs 80 gallons, on an average, yield an equal weight—or 100 pounds of the water yield from 15 to 18 pounds of salt.

It is found, however, that, by constant pumping, the water gradually becomes weaker in salt, and unfit for profitable use; when either the water in that well must be allowed to rest for a time, or new borings must be put down. This shows that the comparatively fresh-water from above, in percolating through the strata of the salt-group to the bottom of the valley in which the drift lies, becomes impregnated with salt contained in these strata; but no facts have yet been observed which justify any confident opinion as to the actual existence of beds of solid rock-salt in this formation.

The quantity of water pumped up and consumed in the summer and autumn, when the salt-manufacture is most active, amounts to about 2,000,000 of gallons a-day, yielding about 35,000 bushels of salt, weighing 56 pounds each. The total annual production of salt in 1848 was 4,700,000 bushels. The importance of this production of salt, in reference to the consumption and wants, not only of the State of New York, but of the whole United States, will appear, on comparing the above quantity with the importations of foreign salt into the port of New York, and into those of the whole Union respectively, in 1848—and with the whole quantity consumed in the State of New York and by the inhabitants of the United States. Thus there was

Manufactured at Syracuse,	.	4,737,126 bushels.
Imported at New York,	.	1,882,473 ...
Imported into the United States,		8,967,600 ...

so that the manufacture at Syracuse is equal to one-half of the entire foreign importation into the States.

As to the relation of the quantity manufactured to the actual wants of the State, it was shown some years ago by Mr Meriam, that the quantity of salt consumed in the United States was at the rate of three pecks (42 lbs.) to each individual. Assuming the population of the State of New York to be 3,000,000, and that of the whole Union 20,000,000, in round numbers, the whole consumption will be—

For the State of New York,	.	2,250,000 bushels.
For the United States,	.	15,000,000 ...

Hence the State of New York makes much more salt than is necessary for its own wants; so that, while it imports at New York nearly one-half of its own consumption, it on the other hand exports, by way of the lakes, to Canada and the Western States, about 3,500,000 bushels.

The large alleged individual consumption of salt in the United States is worthy of attention. In ordinary families in Great Britain the quantity of salt used for domestic purposes is about 12 lb. for each individual; and if as much more be used for all other purposes, the consumption ought to amount to 24 lb. a-head, or, in all, to less than 10,000,000 of bushels. But in the United States the consumption is estimated at three pecks, or 42 lb., for each individual—which large allowance, considering there are few chemical manufactories to eat it up, and little is employed for agricultural purposes, appears to imply either a large waste, or an outlet for it which does not exist in this country. It is possible that the large quantities of salt provision which are prepared for home consumption and for export—"the immense packing business of the West,"* as it is called—may be a main cause of the increased proportional use of salt in North America, if the estimate be a correct one.

The salt-springs of Onondaga are the property of the State; and by an article of the Constitution, they and the lands necessary for the manufacture of the salt can never be alienated. The wells are sunk and the water pumped up into reservoirs at the expense of the State, and thence distributed to the various manufactories, for a charge of one cent upon each bushel of salt manufactured. Until the year 1846, the duty levied as a charge for the water amounted to six cents per bushel, and a clear annual revenue was obtained from the springs of about 200,000 dollars.

* In 1848, 34,000,000 of pounds of salted meats, and nearly 3,000,000 of pounds of butter, were exported from the States.

In that year, however, the duty was reduced to one cent, and they now yield a net revenue of about 30,000 dollars a-year. But the production has been largely increased, and the State is greatly benefited in other ways by the reduction. About twelve cents a bushel is now the market price of the Syracuse salt.

The salt is manufactured in three ways: First, by solar evaporation. This is conducted in large wooden vats 18½ feet square, and about 12 inches deep, provided with movable lids, by which they can be easily covered in rainy weather. These vats are arranged in long rows communicating with each, and at different levels, so that the water which is put in at one end can be made to flow forward as its concentration advances. The iron is the first constituent of the water which is deposited; after that the gypsum, which forms beautiful small crystals on the bottom and sides of the vats; and then the common salt, in large coarse crystals like those of bay salt—the deliquescent magnesia and lime salts remaining in the waste mother liquors.

The second mode is similar to that followed on our coasts in extracting salt from sea-water, by evaporation in large shallow iron pans. The third method is by boiling the water in deep iron pots, or kettles, as they are called, of which forty, built up in two parallel rows, form what is called a block. These kettles are supplied with water from the private cistern of the establishment in which the brine has been purified, by standing some time, or by the addition of quicklime, which carries down the iron, magnesia, and some other impurities. As the water boils away, gypsum first falls in the kettle, and is continually ladled out as it collects at the bottom; and as the liquor concentrates, the salt falls in the form of a fine white powder, which is lifted out, set to drain, and dried. It is then ready for market. This is the favourite and quickest mode; but it makes a less pure salt. About

300 acres are covered with vats for solar evaporation; and these, in 1848, yielded 713,000 bushels, while only 75 acres are occupied by the *blocks*, which manufactured upwards of 4,000,000 bushels of fine salt. About one half of the coarse salt is crushed or ground, and sold under the name of dairy salt, being preferred for dairy purposes.

In the order of deposition of the several ingredients of the natural brine in the solar vats, Mr Vanuxem finds a resemblance to, and an explanation of, the mode and order of deposition of the several members of the Onondaga salt-group. "This group shows first a thick mass coloured red with iron, being its red shale." This corresponds with the oxide of iron first deposited in the vats. "Above the red shales are the gypseous masses, towards the upper part of which are the salt cavities"—(that is, the hopper-shaped cavities in which crystals of salt are supposed formerly to have existed.) "Above the whole of these deposits is the sulphate of magnesia, its (former) existence there being manifested by the needle-form cavities" in the rock.* This explanation is very natural, and not void of beauty. It may require some modifications to adapt it to the local phenomena in detail—such as the occurrence of the green and blue shales, the limestones, and the calcareous marls. It is, however, neither an unnatural nor unlikely general representation of the probable cause of some of the special chemical characters exhibited by the several successive beds.

A more refined examination of the salts successively deposited in the open wooden troughs in which the water is exposed to spontaneous evaporation might lead to interesting results of another kind. The brine most probably contains both iodine and bromine; and it is possible that, at a certain stage of the evaporation, the saline compounds of

* *Geology of the Third District*, p. 109.

these substances (iodides and bromides) may either be deposited more abundantly, or retained in solution in larger proportion than the common salt, and might in consequence, in such large operations, be largely, easily, and profitably extracted. I suggest the examination of this point to my chemical friends in the State of New York especially, as a very likely source of abundant supplies of bromine.

CHAPTER VII.

Railway to Buffalo.—The Americans a *clever* people.—Incorrectness of speech on both sides of the Atlantic.—Joe Smith, founder of the Mormons.—Outline of his history while in the State of New York.—His removal to Missouri, to Ohio, and Illinois successively. His imprisonment and death.—Rapid progress of his sect.—New State of Utah, on the Salt Lake.—Outline and character of the book of Mormon.—Its claims as an American revelation.—Canandagua.—City of Rochester; its rapid rise.—Genesee flour.—Money-value of farms on the Genesee River.—Profits of farming in this valley.—Mr Wadsworth's farms and farming.—Rent, and rotation on his land.—Capital of farmers on this estate.—Inducements to invest money in land in New York State.—Sowing and reaping of wheat.—Relative values of rural produce and of human labour.—Average produce of the Genesee country.—New York does not produce wheat enough for its own consumption.—North-east America not a dangerous competitor in the English wheat market.—Upper Canada might for a time successfully compete with the English farmer.—Duty upon Canadian wheat in the United States ports.—Expected effects of a repeal of this duty.—Made an argument for annexation.—Importance of the direct trade to Europe by the St Lawrence.—Erie Canal; its length, and that of its branches.—Amount of traffic and revenue.—Number of emigrants from different countries.—Cost of passage from New York to Lake Erie.—Influence of the New England States on the development of the new States.—Democratic party in the United States.—Principles of the Old Hunkers and the Barnburners.—Forest and half cleared land on approaching Buffalo.—Speculators in land.

SEPT. 15.—At seven in the morning I was at the railway station to take my departure for Buffalo. I here encountered one of those embarrassing *mes-entendres* which are unavoidable when travelling among people who *will* use old words in new senses. I was introduced to a gentlemanly-looking physician, who followed up his question of how I liked the country, with the question, "Don't you find us a very *clever* people?" This was a thrust so decidedly home that I could not believe he meant what he asked. He looked also perfectly innocent, but evidently expecting an answer. As I could not conscientiously say yes, I hesitatingly said what had more than once occurred to me in passing through the States—"At least you think yourselves so." But the instant the words had escaped me, I apologised for my rude speech, recollecting that, only two or three days before, an American lady had remarked to me that this word, in the States, is often used in the sense of "good-natured, or ready to oblige." That the people of the United States, wherever I have been, are clever in this sense, I can honestly say. I added, therefore, *avec empressement*, "I understood your question wrongly; in your sense of the word, you *are* a clever people." My new acquaintance felt the situation as much as I did, and explained his question as a bit of local slang. A stranger ought not willingly to give offence to a people among whom he is travelling, nor to say what is likely to hurt their feelings; but this was a case where the fault was on the side of those who are not careful to maintain the fountains of speech in their primitive purity.

Were we, however, to criticise the home speech of the midddle classes among ourselves, as we feel inclined to notice peculiarities abroad, we should find many more instances of incorrectness than we are generally aware of. Just before I left home, a lady of my acquaintance, the daughter and sister of a clergyman, being asked of her sister's health, answered that "she was very shabby;" and here, in this western New York, I have been talking to an Englishman, on his travels like myself, who tells me that the boat on Lake Ontario does not sail *whilst* to-morrow. Though these expressions are quite intelligible, yet they afford some ground for the opinion which a Yankee will occasionally express, that his countrymen talk English quite as well as ourselves.

Of course, I do not allude to provincial dialects, which have not yet had time to spring up in the States, but which, as with us, will gradually arise out of the different nationalities settled in different districts.

The distance by rail to Buffalo is 180 miles, which we took fully twelve hours to accomplish. The whole day's ride was along the belt of wheat-country of which I have already spoken, though by no means in a straight line, or always on its richest and most improved parts.

In the future history of mankind, if present appearances are to be trusted, the counties of Wayne and Ontario, through which we passed in the early part of the day, are likely to derive an interest and importance, in the eyes of a numerous body of people, from a circumstance wholly unconnected either with their social progress, or with their natural productions or capabilities. In these counties lie the scenes of the early passages in the life of Joe Smith, the founder of the sect of the Mormons.

Born in December 1805, in Sharon, Windsor County, State of Vermont, he removed with his father, about 1815, to a small farm in Palmyra, Wayne County, New York, and assisted him on the farm till 1826. He received little education, read indifferently, wrote and spelt badly, knew little of arithmetic, and, in all other branches of learning he was, to the day of his death, exceedingly ignorant.

His own account of his religious progress is, that as early as fifteen years of age he began to have serious ideas regarding the future state, that he got into occasional ecstasies, and that in 1823, during one of these ecstasies, he was visited by an angel, who told him that his sins were forgiven—that the time was at hand when the gospel in its fulness was to be preached to all nations—that the American Indians were a remnant of Israel, who, when they first emigrated to America, were an enlightened people, possessing a knowledge of the true God, and enjoying his favour—that the prophets and inspired writers among them had kept a history or record of their proceedings—that these records were safely deposited—and that, if faithful, he was to be the favoured instrument for bringing them to light.

On the following day, according to instructions from the angel, he went to a hill which he calls Cumorah, in Palmyra township, Wayne County, and there, in a stone chest, after a little digging, he saw the records; but it was not till four years after, in September 1827, that "the angel of the Lord delivered the records into his hands."

"These records were engraved on plates which had the appearance of gold, were seven by eight inches in size, and thinner than common tin, and were covered on both sides with Egyptian characters, small and beautifully engraved. They were bound together in a volume like the leaves of a book, and were fastened at one edge with three rings running through the whole. The volume was about six inches in thickness, bore many marks of antiquity, and part of it was sealed. With the records was found a curious instrument, called by the ancients Urim and Thummim, which consisted of two transparent stones, clear as crystal, and set in two rims of a bow"—a pair of pebble spectacles, in other words, or "helps to read" unknown tongues.

The report of his discovery having got abroad, his house was beset, he was mobbed, and his life was endangered by persons who wished to possess themselves of the plates. He therefore packed up his goods, concealed the plates *in a barrel of beans*, and proceeded across the country to the northern part of Pennsylvania, near the Susquehannah River, where his father-in-law resided. Here, "by the gift and power of God, through the means of the Urim and Thummim, he began to translate the record, and, being a poor writer, he employed a scribe to write the translation as it came from his mouth." In 1830 a large edition of the *Book of Mormon* was published. It professes to be an abridgment of the records made by the prophet Mormon of the people of the Nephites, and left to his son Moroni to finish. It is regarded by the Latter-day Saints with the same veneration as the New Testament is among Christians.

The Church of the Latter-day Saints was organised on the 6th of April 1830, at Manchester, in Ontario County, New York. Its numbers at first were few, but they rapidly increased, and in 1833 removed to the State of Missouri, and purchased a large tract of land in Jackson County. Here their neighbours tarred and feathered some, killed others, and compelled the whole to remove. They then established themselves in Clay County, in the same State, but on the opposite side of the river. From this place again, in 1835, they removed eastward to the State of Ohio, settled at Kirtland, in Gauga County, about twenty miles from Cleveland, and began to build a temple, upon which sixty-thousand dollars were expended. At Kirtland a bank was incorporated by Joe and his friends, property was bought with its notes, and settled upon the Saints, after which the bank failed—as many others did about the same time—and Ohio became too hot for the Mormons. Again, therefore, the Prophet, his apostles, and a great body of the Saints, left their home and temple, went westward a second time to the State of Missouri, purchased a large tract of land in Caldwell County, in Missouri, and built the city of the

"Far West." Here difficulties soon beset them, and in August 1838 became so serious that the military were called in; and the Mormons were finally driven, unjustly, harshly, and oppressively, by force of arms, from the State of Missouri, and sought protection in the State of Illinois, on the eastern bank of the Mississippi. They were well received in this State, and after wandering for some time—while their leader, Joe Smith, was in jail—they bought a beautiful tract of land in Hancock County, and, in the spring of 1840, began to build the city and temple of Nauvoo. The legislature of Illinois at first passed an act giving great, and, probably, injudicious privileges to this city, which, in 1844, was already the largest in the State, and contained a population of about twenty thousand souls. The temple, too, was of great size and magnificence—being 128 feet long and 77 feet high, and stood on an elevated situation, from which it was visible to a distance of 25 or 30 miles. In the interior was an immense baptismal font, in imitation of the brazen sea of Solomon—"a stone reservoir, resting upon the backs of twelve oxen, also cut out of stone, and as large as life."

But persecution followed them to Illinois, provoked in some degree, no doubt, by their own behaviour, especially in making and carrying into effect city ordinances, which were contrary to the laws of the State. The people of the adjoining townships rose in arms, and were joined by numbers of the old enemies of the Mormons from Missouri. The militia were called out; and, to prevent further evils, Joe Smith and one of his brothers, with several other influential Saints, on an assurance of safety and protection from the governor of the State, were induced to surrender themselves for trial in respect of the charges brought against them, and were conducted to prison. Here they were inconsiderately left by the Governor, on the following day, under a guard of seven or eight men. These were overpowered the same afternoon by an armed mob, who killed Joe Smith and his brother, and then made their escape. After this, the Mormons remained a short time longer in the Holy City; but the wound was too deep seated to admit of permanent quiet on either part, and they were at last driven out by force, and compelled to abandon or sacrifice their property. Such as escaped this last persecution, after traversing the boundless prairies, the deserts of the Far West, and the Rocky Mountains, appear at last to have found a resting-place near the Great Salt Lake in Oregon. They are increasing faster since this last catastrophe than ever; and are daily receiving large accessions of new members from Europe, especially from Great Britain. They form the nucleus of the new State of Utah, this year erected into a territory of the United States, and likely, in the next session of Congress, to be elevated to the dignity of an independent State. So rapidly has persecution helped on this offspring of ignorance, and tended to give a permanent establishment, and a bright future, to a system, not simply of pure invention, but of blasphemous impiety, and folly the most insane.

The *Book of Mormon*, which is the written guide of this new sect, consists of a series of professedly historical books—a desultory and feeble imitation of the Jewish chronicles and prophetical books—in which, for the poetry and warnings of the ancient prophets, are substituted a succession of unconnected rhapsodies and repetitions, such as might form the perorations of ranting addresses by a field preacher, to a very ignorant audience.

The book, in the edition I possess, consists in all of 634 pages, of which the first 580 contain the history of a fictitious personage called Lehi and that of his descendants for the space of a thousand years.

This Lehi, a descendant of Joseph the son of Jacob, with his family left Jerusalem in the beginning of the reign of Zedekiah, six hundred years before Christ, and, passing the Red Sea, journeyed eastward for eight years till they reached the shore of a wide sea. There they built a ship, and, embarking, were carried at length to the promised land, where they settled and multiplied. Among the sons of Lehi one was called Laman and another Nephi. The former was wicked, and a disbeliever in the law of Moses and the prophets; the latter, obedient and faithful, and a believer in the coming of Christ. Under the leadership of these two opposing brothers, the rest of the family and their descendants arranged themselves, forming the Lamanites and the Nephites, between whom wars and perpetual hostilities arose. The Lamanites were idle hunters, living in tents, eating raw flesh, and having only a girdle round their loins. The skin of Laman and his followers became black; while that of Nephi and his people, who tilled the land, retained its original whiteness. As with the Jews, the Nephites were successful when they were obedient to the law; and, when they fell away to disobedience and wickedness, the Lamanites had the better, and put many to death. At the end of about four hundred years, a portion of the righteous Nephites under Mosiah, having left their land, travelled far across the wilderness, and discovered the city of Zarahemla, which was peopled by the descendants of a colony of Jews who had wandered from Jerusalem when King Zedekiah was carried away captive to Babylon, twelve years after the emigration of Lehi. But they were heathens, possessed no copy of the law, and had corrupted their language. They received the Nephites warmly, however, learned their language, and gladly accepted the law of Moses.

This occupies 158 pages. The history of the next two hundred years follows this new people, and that of occasional converts from the Lamanites—called still by the general name of Nephites in their struggles with the Lamanites, and the alternations of defeat and success which accompany disobedience or the contrary. This occupies several books, and brings us to the 486th page, and the period of the birth of Christ. This event is signified to the people of Zarahemla by a great light, which made the night as light as mid-day. And thirty-three years after there was darkness for three days, and thunderings and earthquakes, and the destruction of cities and people. This was a sign of the crucifixion. Soon after this, Christ himself appears to this people of Zarahemla in America, repeats to them in long addresses the substance of his numerous sayings and discourses, as recorded by the apostles; chooses twelve to go forth and preach and baptize; and then disappears. On occasion of a great baptizing by the apostles, however, he appears again; imparts the Holy Spirit to all, makes long discourses, and disappears. And, finally, to the apostles themselves he appears a third time; and addresses them in ill-assorted extracts and paraphrases of his New Testament sayings.

The account of these visits of our Saviour to the American Nephites, and of his sayings, occupies about 48 pages. For about 400 years, the Christian doctrine and church thus planted among the Nephites had various fortune; increasing at first, and prospering, but, as corruptions came in, encountering adversity. The Lamanites were still their fierce enemies; and as wickedness and corrupt doctrine began to prevail among the Christians, the Lamanites gained more advantages. It would appear, from Joe Smith's descriptions, that he means the war to have begun at the Isthmus of Darien—where the Nephites were settled, and occupied the country to the north, while the Lamanites lived south of the isthmus. From the isthmus the Nephites were gradually driven towards the east, till finally, at the hill of Cumorah, near Palmyra, in Wayne County, western New York, the last battle was fought, in which, with the loss of 230,000 fighting men, the Nephites were exterminated! Among the very few survivors was Moroni, the last of the scribes, who deposited in this hill the metal plates which the virtuous Joe Smith was selected to receive from the hands of the angel. This occupies to the 580th page.

But now, in the Book of Ether, which follows, Joe becomes more bold, and goes back to the tower of Babel for another tribe of fair people, whom he brings over and settles in America. At the confusion of the languages, Ether and his brethren journeyed to the great sea, and, after a sojourn of four years on the shore, built boats under the Divine direction, water-tight, and covered over like walnuts, with a bright stone in each end to give light! And when they had embarked in their tight boats, a strong wind arose, blowing towards the promised land, and for 344 days it blew them along the water, till they arrived safe at the shore. Here, like the sons of Lehi, they increased and prospered, and had kings and prophets and wars, and were split into parties, who fought with each other. Finally, Shiz rose in rebellion against Coriantumr, the last king, and they fought with alternate success, till two millions of mighty men, with their wives and children, had been slain! And, after this, all the people were gathered either on the one side or the other, and fought for many days, till only Coriantumr alone remained alive!

This foolish history is written with the professedly religious purpose of showing the punishment from the hand of God which wicked behaviour certainly entails; and, with some trifling moralities of Moroni, completes the *Book of Mormon.*

Joseph Smith does not affect in this gospel of his to bring in any new doctrine, or to supersede the Bible, but to restore "many plain and precious things which have been taken away from the first book by the abominable church, the Mother of Harlots." It is full of sillinesses, follies, and anachronisms; but I have not discovered, in my cursory review, any of the immoralities or positive licentiousness which he himself practised, directly inculcated. He teaches faith in Christ, human depravity, the power of the Holy Ghost, the doctrine of the Trinity, of the atonement, and of salvation only through Christ. He recommends the sacraments of baptism and the Lord's Supper; and, whatever his own conduct and that of his people may be, certainly in his book prohibits polygamy and priestcraft.

The wickedness of his book consists in its being a lie from beginning to end, and of himself in being throughout an impostor. Pretending to be a "seer"—which, he says, is greater than a prophet—he puts into the hands of his followers a work of pure invention as a religious guide inspired by God, and which, among his followers, is to take the place of the Bible. Though an ignorant man, he was possessed of much shrewdness. He courted persecution, though he hoped to profit, not to die by it. Unfortunately, his enemies, by their inconsiderate persecution, have made him a martyr for his opinions, and have given a stability to his sect which nothing may now be able to shake. It was urged by Smith himself that the New World was as deserving of a direct revelation as the Old; and his disciples press upon their hearers that, as an *American revelation*, this system has peculiar claims upon their regard and acceptance. The feeling of nationality being thus connected with the new sect, weak-minded native-born Americans might be swayed by patriotic motives in connecting themselves with it. But it is mortifying to learn that most numerous accessions are being made to the body in their new home by converts proceeding from this country.* Under the name of the "Latter-day Saints," professing the doctrines of the gospel, the delusions of the system are hidden from the masses by the emissaries who have been despatched into various countries to recruit their numbers among the ignorant and devoutly-inclined lovers of novelty. Who can tell what two centuries may do in the way of giving a historical position to this rising heresy?

Leaving behind us the townships of Palmyra and Manchester, the scene of Joseph's first transactions, and of the last battles of his heroes, who seem to have fought very much like the Kilkenny cats, and Canandagua, a pretty town of one long street, running down to the lake of the same name, we rapidly approached the city of Rochester, on the falls of the Genesee River. The valley of this river is celebrated for its production of wheat, and for the first-class flour into which its grain is converted by the Rochester millers. On the falls of the Genesee river have been established the numerous mills and factories for which the city of Rochester is famous. In 1812, only ten houses stood where, in 1850, a city with 25,000 inhabitants had already arisen. The flour-mills, which are driven by the great water-power, are the chief distinction of Rochester. Some of these consume upwards of 2000 bushels of wheat a-day, and produce, in the same time, 500 barrels of flour. But it has cotton-mills, carpet factories, paper-mills, machine and engine shops, plough, thrashing-machine, and other agricultural implement manufactories, and ship-yards.

* It has been recently stated that the Mormon emigration from Liverpool alone, up to the present year, has been 13,500, and that they have, on the whole, been superior to and better provided than the other classes of emigrants. Of course, many more of this sect must have emigrated from other ports, and many even from the port of Liverpool, whose faith and ultimate destination was not known.

Its literary tastes may be judged of from the fact that, besides common and grammar schools, a Protestant university has just been opened; that the Roman Catholics have a seminary, which they talk of converting into a university; and that, besides general libraries, it has a law library belonging to the State, already containing 3000 volumes.

Not only rapid prosperity, but an active and energetic population of *men born elsewhere*, is implied in all this. Local advantages gave rise to Rochester, and will advance it very much farther. But whatever be the local position, an introduction of new blood—of men who will look upon old and familiar things with new eyes—always helps a place forward. Progress, therefore, cannot fail to be rapid, when, in addition to manifest physical advantages, still imperfectly developed, the blood of the whole city is *new*, untied and untrammelled by old notions, or hampered by forefather prejudices.

Rochester is the proper home of the celebrated Genesee flour. But the Genesee flour of the New York merchants is something like the Wallsend coals of the London Coal Exchange. It is the brand or designation of a superior quality of flour, which has obtained a name in the New York and Atlantic markets, and which is made now from wheat grown in various western localities. The quantity of flour manufactured at Rochester, in 1848, was about 700,000 barrels, requiring upwards of 3,000,000 of bushels of wheat. Of the yearly increasing quantity thus ground up by the Rochester millers, a large proportion is annually brought from other States by way of the Erie Canal, the railroad, and the lakes.

The Genesee Valley, which gives its name to the finest samples of flour, produces wheat of the best quality. The soil on which it is grown is for the most part a rich drift clay—the ruins of the Onondaga salt-group—intermixed with fragments of the Niagara and Clinton limestones. A very comfortable race of farmers is located in this valley. The richest bottom or intervale land, cut for hay or kept for grazing, is worth 120 dollars, or £26 an acre. The upland, the mixed clay and limestone gravel land, of which I have already spoken, when sold in farms of 100 to 150 acres—the usual size on this river—brings from 35 to 70 dollars, according to the value of the buildings that are upon it. The bottoms, when ploughed up and sown to wheat, are liable to rust; but the uplands yield very certain crops of 15 to 20 bushels an acre. On crops of 15 bushels the farmers of all this wheat-region can live very well.

Land, of which a man with a good team will plough 1¼ to 1½ acres a-day, costs six dollars an acre to cultivate, including seed, and 3½ more to harvest and thrash. Fifteen bushels, at 1 to 1⅛ dollars, (4s. 4d. to 4s. 10d.,) give a return of 15 to 17 dollars, leaving a profit of about six dollars, or 26s. an acre, for landlord and tenant's remuneration, and for interest of capital invested in farming stock. That this calculation is near the truth, is shown by the rate at which the average land, producing 16 to 18 bushels, is occasionally let, where it suits parties to make such an arrangement. In these cases, 7 to 7½ bushels of wheat an acre are paid for the use of the land. In taking a farm at such a rent as this —half the produce—the tenant makes a sacrifice for the purpose of obtaining an outlet for superfluous home labour. Our small farmers of 50 or 100 acres, who cultivate with their own families, do the same when they consent to pay rents which leave them, out of the produce of their farm, at the end of the year, less than the usual wages of the labour expended upon it, for the accommodation and comfort of being their own masters, and of living and working together.

But all through this wheat-district much land is let

upon shares, the cultivator paying one-half of the clear produce, with such details as the parties may arrange. As in New Brunswick, this system is more popular than that of a fixed, and especially a money rent.

I had subsequently an opportunity of meeting Mr Wadsworth, one of the largest landowners in the State, possessing a large tract of the wheat-land in the Genesee Valley. This gentleman himself farms 1000 acres, and clears from 3 to 7½ per cent on the whole capital employed, including the market value of the land, and of the buildings and stock upon it. For a gentleman farmer, this would be a very fair return; but it is scarcely enough in a country where land gives no political and little social influence, and where, by lending his money, and doing nothing, a man can obtain 7 per cent certain.

Mr Wadsworth informed me that the system of renting farms is not unpopular in his district; that his farms used to be let nominally on shares, but in reality at a fixed grain rent. The produce was estimated at 18 bushels of wheat an acre, and he took one-third, or 6 bushels, as the rent. Latterly he has been taking 8 bushels, and the farmers pay it readily. The rotation he prescribes is wheat followed by two years clover, cut for hay or eaten off the first year, and eaten off or ploughed in the second. For the wheat land he takes 6 or 8 bushels of grain of the best quality, delivered in kind at a warehouse on the canal, where it is always sure of an immediate and ready market;* for the clover land he takes a money-rent of two or three dollars an acre, as may be fixed by inspection of an agent, every year.

* This is the old Scotch system of corn rents without the averages, which the easy sale of wheat in this locality renders unnecessary. Nothing seems fairer for all parties, more suited to a proper rotation of crops, or less likely to be affected injuriously either for landlord or tenant by changes in corn laws, than that which is still followed where old leases are unexpired. Thus on the Hamilton estate, within 6 miles of Glasgow, a farm of 150 Scottish acres, the lease of which is just

These tillage farms are cultivated by persons who do not usually possess more than £1 an acre of capital; they afford, in fact, an opportunity for persons to begin life who do not possess money enough of their own to buy farms, at least in that neighbourhood. What kind or quality of farming would be looked for, in any of our best districts, from men with such a capital as this?

I may here remark, indeed, as my general impression in regard to the farming of the whole of north-eastern America it was my fortune to visit—that too little capital is employed in cultivating the land. The land itself, and the labour of their families, is nearly all the capital which most of the farmers possess. And if any of them save a hundred dollars, they generally prefer to lend it on mortgage at high interest, or to embark it in some other pursuit which they think will pay better than farming, than to lay it out in bettering their farms, or in establishing a more generous husbandry.

Of the rich grazing land, an acre and a half fattens off a beast which in the lean state will cost £5. Those who hold this land, therefore, require a capital of £3 or £4 an acre to stock it.

With a capital of £1 an acre only, an exhausting system of culture can scarcely fail to be followed—especially as the custom is to remain only from four to eight years, and during this time to save as much as enables the tenant to buy a farm somewhere for himself. Hence the necessity for the strictest adherence to a rotation such as that I have mentioned.

On the whole, under this system of management, Mr Wadsworth calculates that his land yields him five per

expiring, pays a rent of 62 quarters each of wheat, barley, and oats, or 3 bushels of wheat, 3 of barley, and 3 of oats per imperial acre, reckoned at the average prices for the same year between November and February. Eight bushels of wheat per acre is a higher rent than the above nine of mixed crops; but the land of the farm I allude to is not naturally so good as that upon the Genesee.

cent upon its market value in the form of rent, besides which he has the benefit of the continual rise in the value of land, which has added enormously to the value of this property since it came into the hands of his family. But this return is scarcely advantageous enough to present an inducement to moneyed men of the Old World to invest their capital in the purchase of large tracts of land in the New. The possession of this land carries with it little increased consideration, and confers no political influence. On the contrary, in most places it is a cause of jealousy, distrust, and dislike. The feeling is that the living man, and not the dead chattels that he owns, ought to influence the destinies of the country.

The wheat in this Genesee region is sown in the last week of August, or first of September. It thus gets well rooted before the winter sets in, and is not so liable to be thrown out. It begins to grow again about the first of May, and from that time is ripe in about eighty days, being usually reaped in the last week of July. This rapid growth, in Mr Wadsworth's opinion, stands in the way of their raising generally, in this district, crops of wheat so large as ours. The heavy crops which are occasionally reaped, however, are not in favour of this opinion. It is, nevertheless, an interesting subject for experiment. What the land of this region would produce, under skilful culture steadily conducted, is not yet known. If the land were enriched with substances which, while they did not raise so great a rush of straw as to increase the chances of rust, should especially promote the growth and filling of the ear, the true influence of the climate upon this important crop would be more distinctly made out.

An interesting fact connected with the agricultural history of this country is, the change that has taken place in the relative values of human labour and of rural produce. Forty years ago the wages of a man for a year were valued at the price of one yoke of oxen, or of 50 bushels of wheat. Now they are equal to the value of two yoke of oxen, or of 100 bushels of wheat. This does not indicate a corresponding money-rise in the wages of labour. The difference is partly caused by a fall in the market-value of agricultural produce; but, as all other necessaries of life are at least as cheap as they were forty years ago, the condition of the agricultural labourer must have greatly improved. I doubt if, in any part of our own islands, the same can be said in regard to the condition of our agricultural labourers.

What this district of country originally was in fertility may be inferred from what it still is. The average produce per acre of its three most fertile counties, as recently published, is represented by the numbers in the following table:—

	Genesee.	Ontario.	Niagara.
Wheat, . .	16½	16	18
Barley, . .	15	19	19
Oats, . .	23	32	29
Rye, . .	10	9	8½
Buckwheat, .	19	21	17
Indian corn,	25	29	29
Potatoes, .	125	106	110
Turnips, .	105	148	155

Compare the numbers opposite wheat and barley, and it will appear at once that this is eminently a wheat-country; and then consider that, after all the exhausting culture to which it has been so long subjected, it still yields an *average* of from 16 to 18 bushels of wheat per acre, and we shall have an idea of what its natural fertility as a region must originally have been—what, I may say, its fertility will still become, when the exertions of its agricultural societies and of its many spirited improvers shall have produced their destined effect, (I hope I may so speak of it) upon the practical cultivation of its grateful soil.

But, with all the fame and natural capability of this fine western region of New York, the Empire state as a whole does not at present—according to the best information I have been able to obtain—produce wheat enough for the consumption of its own inhabitants. In round numbers, the population of the State of New York is about three millions, while the produce of wheat in bushels is about 15 millions—or at the rate of five bushels to each inhabitant. If, according to our English calculations, eight imperial bushels a-head are necessary to support our people, the New York people, though they may consume a large quantity of Indian corn, can have little surplus left out of five bushels a-head.*

My impression is, therefore, that our British farmers have little to fear from the competition of the wheat-growers of all that part of North America which lies between the Atlantic and the St Lawrence—from Newfoundland on the east, as far west as the head of Lake Ontario. The whole of this region, as to wheat-growing, is more or less in sympathy with the English farmer — struggling against the vigorous competition of the new North-Western States. Wheat is already more costly to raise than in former years, even in the least exhausted portions of this north-eastern region; while the North-Western States, until the rise or growth of a consuming population among themselves, or till the freshness of the virgin soils is rubbed off, can afford to export supplies of wheat at a comparatively nominal price.

I would not be so rash as to say, or so uncharitable as to hope, that the wheat-producing powers of the region east of Lake Erie, and south of the St Lawrence, will never be much greater than it is now; I believe it may become, and I hope the time may soon arrive, when more skill and knowledge shall have forced it to become

* This argument is stronger, if it is also recollected that an imperial quarter is a little more than nine American bushels.

far more productive, as a whole, than it is now. But by that time a larger home population also will have arisen to consume it, while the very introduction of more generous and more artificial forms of culture will add to the cost of production on the whole. So that, even when such better agricultural times arrive in this region, the English farmer will still, in my opinion, have little to fear from this quarter of the American continent. He may find here new Lothians and Norfolks and Lincolnshires, and a reproduction of the best farmers of all these districts—their very sons and grandsons, in fact, settled on American farms; but I do not expect he will ever, in ordinary seasons, have reason to look upon them with bitter feelings, as so rivalling him in the British wheat market as to lessen his own home profits, or seriously to diminish the comforts of his family.

Were Upper Canada seriously to turn her thoughts to the reasonable development of her material interests, she ought to become, for a while, the most formidable American competitor of the English farmer, in the home markets.

The duty upon Canadian wheat on entering the States is 20 per cent. This, as is well known, is loudly complained of, and it is desirable that it should be removed, on the broad principle of abolishing restrictions upon commerce—if not for those special reasons which our North American colonies put forward most strongly. But notwithstanding this duty, the Rochester millers cross the lake, buy the Canadian wheat at Toronto, pay the freight across, and the 20 per cent duty, and after all, make a profit on their flour in New York. Is this first-class flour with the Genesee brand made for the home market only—to suit a special American taste—and is a higher price given to the Rochester millers on this account? If so, one could understand why, with the view of manufacturing for this special market, the Rochester millers could afford to pay this premium for Canadian wheat of equal quality with that of western New York, and grown upon similar soils. But such is not the case. Abundance of this flour, if I am not misinformed, comes to Liverpool, and is sold in the English, and I believe also in the New Brunswick and Nova Scotia markets. What a premium there is here, therefore, upon the expenditure of Canadian enterprise!

At the period of my visit to the neighbourhood of Rochester, wheat was selling in the Toronto market at 75 to 80 cents a bushel; while at Rochester, the same qualities brought 106 to 112½ cents—being a difference of about 25 per cent. In six hours a packet can cross from Toronto to Rochester and bring grain at 2 cents a bushel; and it is the opinion of the Upper Canadians, that if they had reciprocity with the States, either by annexation or otherwise, their wheat would rise to the full amount of the duty now paid on importation to the States. The Rochester and Oswego millers, on the other hand, are favourable to reciprocity, because they think that a large proportion of the duty now paid, would come into their pockets; while the New York farmers are naturally averse to the introduction of Canadian wheat duty-free, to compete with, and reduce the price of their own, at the season of the year in which it is in greatest request.

I shall return to this subject hereafter, when I speak of my observations in Canada. I will only observe, in this place, that the expectations of the Canadians from this reciprocity are in my opinion greatly exaggerated. The differential duty ought to be removed if possible, on the ground that all such obstructions stand in the way of the full and natural development of the resources of great countries, and create artificial interests, as they have done among us, which it is very difficult afterwards to do away with. But the policy of Canada is to encourage direct communication with Europe, and to endeavour to perfect itself to make more use of the navigation of the St Lawrence. Western America is now beginning to feel the importance of this river, which is the great natural outlet, and must become a main commercial outlet for the growing productions of the new North-Western States. The Erie Canal is already far too limited for the traffic which at present comes to it, and the St Lawrence presents facilities which the canal does not possess; and of these facilities numerous parties in the States are already beginning to avail themselves. Instead of making sacrifices, therefore, with the view of obtaining a reciprocity, the benefits of which would be at least as great to New England as they would be to Canada, let the Canadian wheat and flour be sent direct to Liverpool; and in this way the Canadian merchant and miller may put in their pockets the duties and profits they are now made to pay to other parties as they pass through the cities of Rochester and New York to the same final destination.

This great canal, the sole main artery along which—until the railway from Albany to Buffalo was formed—the traffic of the west and of the upper lakes found its way to the Atlantic, is not only very creditable to the enterprise and foresight of the people of New York State, but has been a source of great wealth to them. It has also proved the means of developing, in a most rapid manner, the natural resources of the State, and of creating an extent of commercial and manufacturing industry which, without such a means of easy, lengthened, and economical transit, it might have taken centuries to establish. This canal connects the navigable waters of the Hudson, at Albany and Troy, with the north-eastern part of Lake Erie at Buffalo. It has a length between these extremes of 363 miles, and it is joined in its way by branches, which with their feeders are 273 miles in length. There is, besides, the Lake Champlain Canal of 79 miles, by which the navigation of the lake is connected with that of the river Hudson, making in all 714 miles of canal navigation, and necessary feeders, executed and upheld by the State. In executing this work, a large outlay was for many years incurred, and its projectors had much opposition to encounter, and many struggles with the purse-holders of the State, before it could be brought into a navigable condition. Now it is a source of large direct money-revenue to the State, in addition to the numerous indirect benefits which it otherwise confers. The total number of tons of goods of all sorts conveyed along it, in 1849, was 2,894,732, and their estimated value about 145,000,000 of dollars. The main transit of produce is from west to east, as we should naturally suppose. Thus, in 1849, the tons of wheat and flour conveyed eastward to the Hudson were 434,444, while there were conveyed westward to Buffalo, on the way to other States, only 67,966 tons of the same articles. Of the whole traffic, also, of all kinds, 1,266,000 tons came eastward to the Hudson, while only 315,000 tons went westward. The toll varies from 1 to 5 mills (thousandths of a dollar) for 1000 lbs. per mile, and the revenue in 1849 amounted to 3,250,000 of dollars, of which $2\frac{1}{3}$ millions remained as a clear surplus, after paying all necessary expenses of collection and repairs.

The importance of this canal to the Western States appears from the fact, that of the whole traffic from the west, (1,226,000 tons) which arrived at the Hudson, 768,000 tons had come direct from the Western States. Besides, it is one of the great thoroughfares to the western territory for the poorer classes of emigrants from Europe. For the emigrants who arrive at New York there is a choice of two routes to the west—one by Philadelphia and Pittsburg, the other by Albany and Buffalo. The latter is the cheaper, the easier, and the more direct, and, since the formation of the railway, is also much the quicker route.

The number of emigrants who landed at New York in the years 1848 and 1849 respectively, were

In 1848, . .	189,176
In 1849, . .	220,603

Of these there were from

	1848.	1849.
Ireland, ...	98,061	112,591
England, ...	23,062	28,321
Scotland, ...	6,415	8,840
Wales, ...	1,054	1,782
Total from Great Britain and Ireland, . .	128,592	151,534
Germany, ...	51,973	55,705
Holland, Norway, and Sweden,	2,932	6,754
France, ...	2,734	2,683

A very large majority of the European population, which is flowing to the United States, comes, therefore, from the United Kingdom. Germany sends out more than any other country of continental Europe. In considering the effect of the vast numbers of Irish emigrants upon the population of North America, it is of consequence to notice that of the Teutonic races, including the English, German, Scotch, Dutch, and Scandinavian: there are almost as many as there are of the Celts. The Irish emigrants, also, are by no means all of pure Celtic blood. As a whole, therefore, these emigrants would produce a valuable mixed population were they to be settled indiscriminately, and intermingled by marriage in succeeding generations. This, however, is to a certain extent prevented, by the natural tendency of people of the same country to flock together, and to settle near each other. Thus, as the French preponderate in Lower Canada and Louisiana, the Germans in Pennsylvania and parts of Ohio, the Dutch in some places upon the Hudson River, &c., so the Irish and the Highland Scotch, and even the Norwegians, establish themselves in separate localities, and give a tone to the manners, feelings, and habits, and new words and accents to the language of the townships, counties, or states, in which they choose their homes. This will, no doubt, be found to give peculiarities to the population of the several States, and to modify the temper, and even the legislation, of the Houses of Assembly in each State. But these State differences will disappear in the Federal Congress, and will rarely affect the action or procedure of the Central Government. The greater fire and impatience of one State will be restrained by the coolness and caution of another; and thus, while the warm temperaments of some of the States may prevent national stagnation, they will, it is to be hoped, but seldom prevail to hurry forward the whole Union to hasty and inconsiderate measures.

In regard to the new States which are springing up towards the west and north, it is very interesting to observe how important an influence is exercised by the restless New Englanders upon the establishment among them of political, religious, and educational institutions, and upon the general character and expression of public feeling and sentiment.

The emigrants who go out from Europe—the raw bricks for the new State buildings—are generally poor, and for the most part indifferently educated. Being strangers to the institutions of the country, and to their mode of working, and, above all, being occupied in establishing themselves, the rural settlers have little leisure or inclination to meddle with the direct regulation of public affairs for some years after they have first begun to hew their farms out of the solitary wilderness. The New Englanders come in to do this. The west is an outlet for their superfluous lawyers, their doctors, their ministers of various persuasions, their newspaper editors, their bankers, their merchants, and their pedlars. All the professions and influential positions are filled up by them. They are the movers in all the public measures that are taken in the organisation of State governments, and the establishment of county institutions; and they occupy most of the legislative, executive, and other official situations, by means of which the State affairs are at first carried on. Thus the west presents an inviting field to the ambitious spirits of the east; and through their means the genius and institutions of the New England States are transplanted and diffused, and determine, in a great measure, those of the more westerly portions of the Union.

Near Rochester we passed one of the emigrant trains which every day proceed from Albany and Troy to Buffalo, on their way to the Far West. There were women and children and men of all ages, and the ragged and lively, though often squalid-looking Irish, were mixed up with the more decently clad and graver-looking English, Scotch, and Germans. The fare from New York to Albany by water, and thence to Buffalo by railway, is five dollars a-head, though the poor strangers are liable to much imposition in New York on the part of a set of men called *runners*, who waylay them on landing, and profess to give them information, with the view only of cheating them of their money. Much pains has been taken, however, by the State Emigration Commissioners in New York, with the view of preventing such imposition. It is made unlawful for a tavern or lodging-house keeper to detain the luggage of an emigrant for any debts he may contract; persons are appointed to give information to those who land; and were ordinary prudence to be exercised by European emigrants, very much less opportunity for fraud would be afforded to the swarms of heartless wretches who, in proportion to the population, are probably as numerous at least in New York as in any European city.

Though most of my fellow-passengers were on their way from the fair at Syracuse, the conversation in the cars was more about political differences, conventions, and discussions, than about the proceedings of the show. The *Old Hunkers* and the *Barnburners*—two sections of the democratic party—were holding many meetings throughout the country, with the view of bringing about a union, as their differences had been a source of great advantage to the Whigs at the recent federal and state elections. In England, to be a democrat still implies a position at the very front of the movement party, and a desire to hasten forward political changes, irrespective of season or expediency. But among the American democrats there is a Conservative and a Radical party. The former, who desire to restrain "the amazing violence of the popular spirit," are nick-named by their democratic adversaries the "Old Hunkers;" the latter, who profess to have in their hearts "sworn eternal hostility against every form of tyranny over the mind of man," are stigmatised as *Barnburners*, but call themselves the "*Young Democracy*," or the "Progressive Young Democracy." The *New York Tribune*, in reference to the origin of the names themselves, says that the name Hunkers "was intended to indicate that those on whom it was conferred had an appetite for a large *hunk** of the 'spoils,' though we never could discover that they were peculiar in *that*. On the other hand, the Barnburners were so named, in allusion to the story of an old Dutchman who relieved himself of rats by burning his barns which they infested, just like exterminating all banks and corporations, to root out the abuses connected therewith." It is alleged against the Barnburners, that

* *Hunk* or *hunch* is a large slice or piece—as, a *hunk* of bread and cheese.

their recent conduct in reference to the slavery question, as the supporters of General Cass, the pro-slavery candidate, is sadly inconsistent with their affected hostility to every form of tyranny. But where men go for the predominance of a party, small considerations regarding consistency will not readily restrain them.

But if the principles of this extreme section of the democratic party in the United States be such as their own organs (the *Ohio Union*, for example,) represent them, they can scarcely be charged with inconsistency in this, or almost any other case. "They believe that the democratic impulses are right, and should be obeyed, not thwarted: they believe in and favour progress, and would not prescribe *a fixed rule in all minor matters for all time*, but would adapt action to the circumstances and exigencies which arise in the progression of events, and to the rights and interests which accompany or result from that progression and its changes." This is virtually surrendering principle to impulse, and giving the reins into the hands of a constantly shifting expediency. If they find it expedient, for party purposes, to oppose the extension of slavery to-day, therefore, it will not, with these professions, be inconsistent to pronounce it expedient to favour that extension to-morrow.

Proceeding from Rochester to Attica, a distance of forty-four miles, in a south-west direction, we again crossed the several geological formations I have already described, and saw much strong wheat-land. Here and there considerable patches of forest remained, and sometimes fields with the stumps standing; and occasionally my memory was refreshed by a more or less extensive burning of the stumps, reminding me of what I had seen so frequently, and on so large a scale, in the forests of New Brunswick. Thirty miles more brought me to Buffalo; and upon this tract the native forest, still untouched, and the log cabin, and the half-cleared land,

and the blackened stumps, and the occasional fellings and burnings, told of our approach to the limits of complete settlement—to the wilderness lands, over which the living tide of redundant European energy is so rapidly diffusing itself.

Along the line of this great thoroughfare in the State of New York, comparatively few emigrants now linger. Farmers, with capital to stock a good farm at home, occasionally find eligible farms to buy, upon which they can comfortably settle, and bring up their families without fear of rent-days or shifting corn-laws. But the mass of movers, who are men of comparatively small means, pass on without inquiring whether or not the State of New York has still any suitable land to sell.

It may at first sight be considered as a remarkable circumstance, indeed, that, in a country so large and so new as the State of New York, containing 46,200 square miles, only 350,000 acres were public property at the beginning of 1849. Of these only 25,000 belonged to the State, 11,000 to the Literature Fund, and 314,000 to the School Fund. But a little inquiry soon shows that when people are flocking in from foreign countries, and lands are for sale at a fixed price, land-speculators will spring up, in whose hands large tracts will accumulate, to be held till a rise in price enables the first purchasers to sell with a profit. It is by land-jobbing, in fact, that the largest fortunes have been made in most of the States. Though this land-jobbing has made it the interest of individuals to use all efforts to turn the tide of emigration in particular directions, and has thus at first more rapidly increased the population of the new States, it has undoubtedly, in the end, the effect of retarding the settlement of a country and the development of its natural resources; and it is one of the internal evils under which our own North American colonies are now to a considerable extent suffering.

CHAPTER VIII.

City of Buffalo; cause of its rapid rise.—Influence of the growth of the Western States on the agriculture of western New York and Upper Canada.—Passage from Buffalo to Chicago in Illinois and Millwaukie in Wisconsin.—Home ideas as to these new States.—Cheap wheat does not imply rich land.—Character of the soils in Michigan.—Average produce of this State, and of its several counties. —Exaggerated statements of the producing and exporting powers of these new States.—Can the export from these new States continue? —Thin sowing of buckwheat.—Quantity of seed-corn per acre sown for the different kinds of grain in the several States of the Union.—Copper mines of Lake Superior.—Immense masses of native copper. —Extent and richness of the deposits.—How they occur.—Ancient Indian workings.—Amusing differences of opinion as to the mode in which the copper has been deposited.—State of Wisconsin.—Popular feeling in regard to the several new States.—Quantity of public land sold in each in 1847.—Short Michigan fever in 1836.—Minnesota, the New England of the West.—Influence of these new States on the future traffic of the St Lawrence. — Wonders of the hog crop of Ohio.—Comparative productiveness of the States of Ohio and New York.—Indian corn the staple of Ohio.—Outlet for this crop in raising pork.—Hogs killed in the several western States.—How they are fed.—"Packing business" at Cincinnati.—How all the parts of the animals are disposed of.—Lard oil exported largely to France to adulterate olive oil.—Amount of the various marketable products of this business at Cincinnati. — Connection of rural economy and manufactures.

BUFFALO, now a city of upwards of 40,000 inhabitants, contained in 1830 only 8,653, and in 1813 was a small village, which in that year was destroyed by fire. Its rise has been rapid, and its future progress is likely to be great; but both are easily intelligible—unavoidable, in fact, in the nature of things. Its position at the termination of the Erie Canal, the monopoly of the carrying trade of the lakes, and the rapid peopling of the Far West—these are the sources of its past progress, and must be the causes of a great increase still to its size, wealth, and importance. It bears, in fact, at one end of the inland water communication of the State, the same relation to the traffic of the wide north-western country as New York at the other end bears to the commerce with Europe.

It is interesting to note the direct and immediate effect which the peopling of a new country has, not only on the rise and prosperity of particular localities, but upon the general wealth, economical value, and forms of husbandry followed in countries which adjoin it.

Thus, in 1838, wheaten flour was shipped at Buffalo *for* the west; and the wheat-region of New York, with that of Upper Canada, were the main sources of its supply. Now, after only twelve years, an enormous supply of wheat and flour is brought *from* the west, along Lake Erie, and shipped upon the Erie Canal for the east, at Buffalo and the adjoining port of Blackrock. Thus, of wheat and flour so shipped, independent of what might be arrested by the way, at Rochester and elsewhere, there arrived at the Hudson River, in—

1846, . . .	264,000 tons.
1847, . . .	398,000 ...
1848, . . .	273,000 ...
1849, . . .	250,000 ...

The value of these articles at Buffalo, in each of these last two years, averaging about 10,000,000 of dollars.

The effect of these large arrivals from the Western States—which were unnaturally stimulated during the years of European famine, as the large number opposite the year 1847 indicates—has been to render wheat less valuable in western New York, to make the wheat culture less remunerative, and to turn the attention of the New York farmers more to grazing and dairy husbandry, fruit culture, and other branches of rural economy, in which they think the north-west will be unable so directly to compete with them.

The nearest of the new North-western States to Buffalo is Michigan, which commences at the other end of Lake Erie. On the arrival of the trains from Albany, steamers, during the summer months, are in waiting to convey emigrants up the lake without delay. At the time of my visit, 500 emigrants a-day were said to leave the Hudson River, and make their way by rail to Buffalo. Along the lake to Detroit, in Michigan—about 250 miles—is, by the quicker boats, 17 hours; from Detroit across the south end of Michigan, by railway, to New Buffalo, 11 hours; and again, by steamboat, across the foot of Lake Michigan, to Chicago in Illinois, is 4 hours; and to Millwaukie, in Wisconsin, about 6 hours more;—in all, about 32 hours to Chicago, (518 miles,) and 42 to Millwaukie.

We are accustomed to attach the idea of great natural productiveness, and of boundless tracts of rich land, to those new States from which come the large supplies of wheat that are annually poured into the port of Buffalo, and which vex the New York State and New England farmers, by their effect upon the prices of the staple article of vegetable food. But a closer examination of these counties undeceives us as to both these points. The power of exporting large quantities of wheat implies neither great natural productiveness, nor permanently rich land, in a district which, from a state of nature, is beginning to be subjected to arable culture.

In Michigan, for example, the geological structure shows that a very large portion of the State is occupied by rocks which belong to the coal measures, and, like the similar rocks in New Brunswick, yield poor soils.

Then the central basin, in which the coal is found, is encircled by a broad zone of those older rocks of which I have spoken under the name of the Portage and Chemung groups, as forming the upland and interior portions of western New York. Among these, thick sandstone rocks occur, which give sandy and stony soils, and dark-coloured shaly rocks, poor in lime, which yield soils on which an inferior forest vegetation naturally springs up, and which, under the influence of culture, yield poorly remunerating returns.

In Michigan *generally*, therefore, the soils ought to be poor — the main exceptions being at its northern extremity, towards the straits of Michilimackinac, which connects Lake Huron with Lake Michigan, and on its south-eastern extremity, where it adjoins the river and lake of St Clair, and the rich lands of the south-western limit of Upper Canada—and such, I believe, is practically found to be the case. The Michigan Central Railroad passes over much poor sandy country. And although extensive tracts of thin poor soil, sparsely studded with open forests of stunted oak, appear to invite the settler by the ease with which they can be cleared, and the first crops put in, yet a few years' trial warns him, at the first favourable opportunity, to shift his location,—sell it, if he can, to a new-comer—and to seek out for a permanent resting-place in a more naturally favoured district.

And yet, such a country as I have described—like the interior uplands of western New York—will give excellent first crops, even of wheat, and will supply, to those who skim the first cream off the country, a large surplus of this grain to send to market.

The correctness of these remarks is proved by a comparison of the actual average produce of the land in this new State of Michigan, with the quantity of wheat and flour it has of late years been able to export. Thus

—according to the Statistical Returns for 1848, published by order of the State legislature in 1849, and for a copy of which I have to express my obligations to the Secretary of the State Agricultural Society, Mr Holmes—it appears that, in 1848, the number of

Acres sown to wheat, was .	465,900
Of bushels produced, was .	4,739,300
And the average, per imperial acre,	$10\frac{1}{5}$ bushels;

or less than 9 bushels, if seed-corn be deducted.

And that this average is not derived from the combinations of extreme numbers, given by very poor and very rich land, appears from the fact, that, of the twenty-nine counties of which the separate averages are deducible from the published returns—

		Bushels per acre.
2	Counties give an average of .	7
3	... .	8
2	... .	9
7	... .	10
6	... .	11
3	... .	12
4	... .	13
1	... .	16
1	... .	18

These last 2 counties being Macomb and St Clair, situated on the fertile south-east portion of the State, to which I have already referred.

And yet the quantity of wheat and flour exported from this State is comparatively large—though I have access to no trustworthy data from which the absolute quantity can be estimated.* Were we to allow to each

* I may give, as an illustration of the very loose, and often exaggerated statements which are put forth regarding these new States, what has been published by authority in regard to Michigan. In the Patent Office Report for 1847, p. 547, a table is given, representing the estimated population, produce in wheat, home consumption of this grain, and surplus for export in each State of the Union in that year. In this table the population of Michigan is taken

of its 400,000 inhabitants 8 bushels of wheat, as Indian corn enters there less into the consumption of the people, there would remain a surplus for exportation of 1,500,000 bushels. Even this is a large export for so young a State, and naturally leads a foreigner to the idea that the country must be very fertile to raise so large a surplus with so small a population. But the real reason for this surplus is, that nearly the whole population is employed in agricultural pursuits, and that wheat is the only grain they produce for which a ready market can be found, and which can be easily exchanged for the manufactured goods, and West or East India produce, which are necessary to their comfort.

A question of great importance to the British and New England wheat-growers here suggests itself—Will the large export of wheat from these new States continue to increase, or are there any reasons why it should by-and-by begin to decrease? So far as I have been able to collect information bearing upon this question, I am decidedly of opinion that, though the quantity of

at 378,000, the produce of wheat at 8,000,000 of bushels! the consumption at 3 bushels per head! and the surplus of 6,890,000 bushels given as remaining for exportation, making Michigan, after Ohio, Virginia, and Pennsylvania, the largest wheat-exporting State in the Union. Again, in the same Patent Office Report for 1848, p. 547, the population is estimated at 400,000, the consumption at 8 bushels per head, and yet a surplus of *four millions of bushels* is supposed to remain for exportation. But in opposition to these calculations are the numbers obtained by actual inquiry in 1848, and published by order of the Michigan legislature in 1849, which show the entire produce of wheat in the State, in 1848, to have been only 4,739,300 bushels.

For such inaccuracies, of course, the compilers of the Reports published at Washington are probably not deserving of blame, so much as their State correspondents, who have a tendency, each of them, to magnify and exaggerate the produce and fertility of his own new region, for the purpose of drawing more men and more capital into it. But their effect, when discovered, is to lessen our faith very much in the other important data which are contained in these very valuable reports, otherwise so creditable to the United States.

wheat and flour exported from these north-western States may continue to increase for a certain limited number of years, it will by-and-by begin to diminish, and will finally in a great measure cease. The reasons for this opinion will be given when I come to review the agricultural changes which the last twenty years have introduced into the husbandry of Lower Canada.

I may however, as a special fact affecting the capabilities of Michigan, here mention, that the long peninsular character of this State, surrounded on all sides by water, mollifies much and equalises its climate, but at the same time makes it exceedingly moist. This, with the peculiar impermeable character of many of its rocks, cover large portions of surface — as is the case in parts of New Brunswick—with bogs and swamps. The whole area of the State is about 36,000,000 acres, and of these 4,500,000, or one-eighth of the whole, are untillable swamps. With an indifferent soil, and a humid climate, this vast extent of constantly cold and exhaling surface must co-operate to diminish the agricultural capabilities of the State as a whole.

In connection with the growth of grain in these new countries, I may advert to a circumstance which is closely related to the subject of a discussion which has for some time interested the agricultural body in Great Britain—that of *thick* or *thin sowing*. I have already alluded, in the case of buckwheat, to the very small quantity of seed—one-half to one bushel an acre—which is found sufficient to secure a good crop of this grain in New Brunswick. In the State of Vermont, from a quarter to a half bushel is the usual quantity of seed sown for this crop ; and in Tennessee one bushel.

I was at first inclined to attribute the success of these small quantities of seed, in the case of buckwheat, to the protection against the attacks of insects which this seed derives from the hard triangular shell with which it is covered; and this may possibly have some influence in securing the growth of a larger proportion of the seeds sown, than in the case of less securely protected kinds of grain. But on comparing together the practice of different portions of the United States, in reference to the quantity of seed-corn of various kinds sown to the acre, we are compelled, I think, to allow to other causes a very material influence in determining how much or how little it will be safest for the farmer to sow. Thus of wheat, barley, oats, rye, rice, and potatoes, the quantity of seed put in per imperial acre, in the different States, so far as I have been able to ascertain, is as follows:—

WHEAT.

	Bushels.		Bushels.
Georgia, .	¾ to 1	Texas, . .	¾
Alabama,	½ ... 2	Illinois,	1 to 1½
Michigan,	1¼ ... 1½	New York,	1¼ ... 2

BARLEY.

	Bushels.		Bushels.
Tennessee,	1	New York,	2 to 3
Michigan,	1½ to 2	New Hampshire,	2½ ... 4
Maine, .	2		

OATS.

	Bushels.		Bushels.
Alabama,	¾ to 1	Michigan,	2 to 3
South Carolina,	1	Vermont, .	3
Georgia, .	1 ... 1½	Iowa, .	2 ... 4
New York,	1½ ... 3	New Hampshire,	3 ... 4

Of these three kinds of grain, it seems to be generally true that the newer States sow less seed than the older, and the more southerly than those which lie more towards the north and the east. As Alabama and Texas are chalk countries, the quantity of calcareous matter in the soils may have an influence in diminishing the quantity of seed-wheat required.

RYE.

	Bushels.		Bushels.
Alabama and Mississippi,	$\frac{1}{2}$	New Jersey and Tennessee	1
Georgia, .	$\frac{3}{4}$	Michigan, .	$1\frac{1}{2}$
South Carolina,	$\frac{5}{8}$	New Hampshire,	1 to 2
		Iowa, . .	2
		Illinois, .	2 to $2\frac{1}{2}$

POTATOES.

	Bushels.		Bushels.
Georgia and Tennessee,	3 to 5	Maine, Connecticut, and Maryland,	10
Mississippi,	4	New York and Ohio,	8 to 20
New Jersey,	$2\frac{1}{2}$ ... 10	New Hampshire, Vermont, and Massachusetts,	10...20

Of these crops, also, the quantity of seed increases as we come north.

Of rice, half a bushel is sown in Alabama, and two bushels in Tennessee; but, in the case of rice and rye, so much depends upon small differences in the soil, and in the case of potatoes upon the variety planted, that safe conclusions cannot be drawn from varying practice, as to these crops.

It is not surprising, though it is economically of much importance, that a still less proportion of seed should be used in planting Indian corn generally, than is found necessary even for buckwheat. Thus the quantities usually put in, are, in—

	Bushel.		Bushel.
South Carolina,	$\frac{1}{10}$	New Hampshire,	$\frac{1}{4}$
Michigan and Georgia,	$\frac{1}{8}$	New York,	$\frac{1}{8}$ to $\frac{3}{8}$
New Jersey, Pennsylvania, and Michigan,	$\frac{1}{8}$ to $\frac{1}{4}$	Ohio, .	$\frac{1}{4}$ ... $\frac{5}{8}$

The differences in regard to this grain are less striking, because it is dropped or dibbled in with the hand; and the above differences in the quantity used depend upon the distances at which the hills are put, and the number of grains dropped into a hill—and these, again, on the height to which the variety grows, and the luxuriance which the soil and climate usually impart to it.

It is a striking fact that, in Alabama, in ordinary husbandry, half a bushel of seed is by very many persons found sufficient to secure a good crop of wheat. Considering the difference of climate, however, it is perhaps as remarkable that, on the low flat new red-sandstone country of north-western Lancashire, called the *Fylde*, where the weather is wet and uncertain in the autumn and winter, less than a bushel of seed should by many persons be found sufficient for their winter wheat, when sown early in October.*

Of the economical circumstances connected in the public mind with the great inland lakes of North America, that which, next to the agriculture of Michigan and the other adjoining States, most materially affects one of our important home interests, is the existence of large deposits of native copper at various places in the States of Michigan and in the British north-western territory. These deposits were known to the native inhabitants of North America, probably many centuries ago, but it is only about ten years since they were re-discovered, and their value proved; and only four or five since the metal from them has been extracted and brought to market.

The most important of these mineral deposits which have yet been discovered are those of the Upper peninsula of Michigan, which separates Lake Michigan from Lake Superior. The mines hitherto opened are chiefly on the Kewenaw peninsula, which juts out into the middle of Lake Superior, and to the south-west of this peninsula, a few miles up the Ontonagon River, which empties into Lake Superior.

The remarkable feature in these mines is the immense sheets, or walls, of native or metallic copper which occur in

* *Royal Agricultural Journal,* vol. x. p. 24.

the veins. It was these masses of solid copper which the native North Americans were in the habit of extracting. Their workings are scattered at intervals over the whole copper region, for a hundred miles, and are sunk twelve or fifteen feet deep. Their working hammers are made of a hard variety of trap—as many as fifty cartloads of them might be collected at the Minnesota mine, near the Ontonagon River! In one of the abandoned pits of this mine, at the depth of twelve feet, a mass of native copper weighing above five tons was found, which had been worked round on all sides, cleared from the rock, and partly raised, by the ancient native miners, and is supposed to have been abandoned because of its great size. Some have imagined that these mining operations must have been conducted by a former race of inhabitants, possessed of more skill than those who were found in the country on the arrival of the Europeans. That the excavations are ancient is proved by the fact that the workings are filled up, and that over them are growing trees which are upwards of a hundred years old. But Dr Charles Jackson states that, with the hammer, tomahawks and scalping-knives have been found; and he conjectures that, as the Indians had no use for the copper but for the manufacture of implements, they had abandoned the mines soon after 1640, when the early missions of the French Jesuits put them in the way of procuring implements of more valuable steel. The succeeding interval of two hundred years would account for all the appearances of antiquity which have as yet been observed.

The Kewenaw Point consists of a series of parallel ridges of trap, which have an E.N.E. direction. The trap contains much iron, and, where it comes in contact with sandstone, assumes the characters of an amygdaloid. In this amygdaloid the veins are rich in metallic copper. In the superincumbent trap they are pinched in to mere films of metal in many places; and in the sandstone the veins, though continuing large, are filled with calc—spar, and other vein-stones—and become poor in valuable mineral matter.

The veins are of two kinds—transverse, which cross the trap ridges in a direction which is N. by 26° to 30° W.; and longitudinal, which run parallel to the ridges and to the strike of the red-sandstone beds. These longitudinal veins, as well as the cross veins, are rich in copper in the amygdaloid only; and though in many places they have the appearance of beds, they are, in the opinion of Dr Jackson, probably true veins. The most successful mine as yet brought into operation is the Cliff Mine, in a cross vein on Kewenaw Point. The vein-stone is prehnite, and the vein widens and becomes richer in copper the deeper it is followed. In 1848 this mine shipped and sold 800 tons, and in 1849 about 930 tons, containing an average of 60 per cent of pure copper. Next to it is the Minnesota Mine, on the Ontonagon River. It is in one of the longitudinal veins or beds, and though only a few hundred tons of metal have yet been shipped, the mine is of great promise. There are many others in progress, from which large supplies are expected, in a few years, to be obtained.

The only form in which the mineral has yet been extracted, in marketable quantity, is in that of metallic copper. When the huge masses and sheets of native metal were first met with in these veins, the extraction of them was found so difficult that it was doubted whether it would be possible economically to remove them. This is now done by clearing the face of the metal, and cutting it out with cold-steel chisels. It occurs in more or less extended sheets, standing on edge, and following the lie of the vein. Several of these are sometimes arranged side by side. They swell out here and there into ellipsoidal forms, so as to be in one place only 6 inches, while in another they are 5 feet in thickness. Dr Jackson states that, at the Cliff Mine, "one mass of pure copper was extracted, when he was surveying the country, which weighed 80 tons; and other masses, probably of equal magnitude, were in process of being uncovered." Mr Trowbridge, in his later Report to the Secretary of State, says that, "in proceeding along the fifth level of the same mine, he passed a mass of copper 125 feet in length, and varying from 1 to 2½ feet in thickness. Its depth was unknown. At one place Captain Jennings (the mine captain) said, 'Here are 100 tons of pure copper in sight.' On the second level we passed another of the same description, &c." In the Minnesota Mine, Mr Hodge describes a sheet which he saw, having a known length of 150 feet, a height of 8 feet, and a thickness, in some places, of 5 feet. What was visible in the overhanging wall of this drift was estimated to contain 250 tons of copper. The poorer part of the vein is blasted out, brought to the surface, heated to redness, quenched in water, stamped, and then washed. By this means the metallic copper is separated, and packed in barrels. This and the chiselled copper are the only forms in which the products of the mines have hitherto been sent to market.

The copper thus obtained, according to Dr Jackson, who first drew the public attention to the value of these deposits of native metal, is pure metallic copper, as dense as the densest hammered copper. It possesses the remarkable peculiarity of being intermixed with variable quantities of metallic silver, not diffused uniformly through the mass, but forming distinct crystals and crystallised masses, scattered through the body of the solid copper. I saw some of these crystals in specimens possessed by Dr Jackson, and he stated to me that some masses are actually *porphyritic*, with metallic silver scattered through them—without, however, forming an alloy with, or rendering impure, the copper itself.

As a matter of economical interest to the copper miners of our own country, I may remark that, in all probability, this part of the United States will, in ten years more, be able to supply, not only a sufficient quantity of copper to meet the demands of the Union, but a surplus also for exportation. The quantity of copper hitherto imported into the United States has amounted to about 5400 tons a-year. The Cliff Mine alone, in 1849, has shipped what is equal to about 560 tons of pure copper.* From what is already known of this copper region, it is fair, I think, to presume that in ten years this quantity will be increased ten times, and that the lake region may be seeking foreign markets for its surplus copper, as the upper Mississippi is already doing for its superabundant lead.

It would be out of place here to discuss the origin of this native copper—the probable mode in which it has been deposited in the veins. It is very amusing, however, to observe in this, as in so many other cases, how men who look at the same phenomena from different points of view see in them arguments for completely opposite views. Dr Jackson remarks, that the metal bears the imprint of crystals of prehnite; that perfect crystals of native copper, sometimes half-a-pound in weight, occur in the prehnite, datholite, calc, spar, and quartz which form the body of the veins; and that the melting-point of silver, which is mixed in crystals and large lumps with the copper, is considerably below that of metallic copper; and these facts, he justly adds, are objections to the igneous origin of these broad sheets of native

* 933 tons containing 60 per cent of copper. The quantity of pure copper extracted at the various smelt-works in Great Britain is about 25,000 tons, one-half of which is from Cornish ores.

metal.* Professor Agassiz, again, looking at the mode in which the mineral deposits occur in the large, thinks the copper all plutonic—that is, thrown up in a melted state; remarks, "that the whole phenomena might easily be reproduced artificially on a small scale;" and adds, "it appears strange to me that so many doubts can still be expressed respecting the origin of the copper about Lake Superior."† So far as I have had the means of forming an opinion upon the chemistry of this subject, I am certainly inclined to differ very widely from the two opinions of Professor Agassiz above quoted—either that the phenomena prove the copper to have been certainly injected in a melted state, or that we really know how to reproduce the phenomena on a small scale by art.

West of Lake Michigan is the new State of Wisconsin, numbering, in 1848, about 220,000 inhabitants. Its principal port, on the south-western part of the lake, and the general destination of emigrants from the east, is Milwaukie. Fifteen years ago Indian skeletons, in rude coffins, might be seen suspended under the trees where this town of 16,000 inhabitants now stands.

Each of the new States has its turn in popular favour, but for the last three or four years the tide of emigration has been setting most strongly towards these North-western States, and especially to Wisconsin. This appears from the extent of the public lands which have been sold during these years in this, compared with the other States of the Union. Thus the Government Return for 1847—the latest to which I have access—gives the following comparative number of acres as sold in the several States during that year:—

* *Proceedings of the American Association for the Advancement of Science*, 1849.—P. 294.

† *Lake Superior: its Physical Character, Vegetation, and Animals.* —P. 427.

	Acres.
Ohio, ...	105,234
Indiana, . . .	230,627
Illinois, . . .	506,802
Missouri, . . .	246,415
Alabama, . . .	146,859
Mississippi, . . .	94,206
Louisiana, . . .	90,694
Michigan, . . .	62,338
Arkansas, . . .	85,448
Wisconsin, . . .	630,575
Iowa, . . .	271,614
Florida, . . .	27,339

This table shows, that to Wisconsin the rush of land-buyers has been greater lately than to any other State, and that Illinois alone approaches it in the quantity of public land disposed of. In 1836, the Michigan fever was the variety then at its height. In that year, four millions of acres of the public lands were disposed of in that State; but 1837 put an end to that excitement, and the sales have since been comparatively small.

North-west of Wisconsin, bounded on the east by Lake Superior, and lying opposite the copper country of Michigan, is the territory—about to become the State—of Minnesota. This district is called by some the New England of the West. It is in a colder latitude than any of the older States, but it is bounded on the west by the Missouri, and is traversed for 900 miles by the Upper Mississippi. These rivers, with their many tributaries, afford abundant facilities for inland navigation. New as this territory is, we already hear of its agricultural societies, its cattle-shows, and its lead-mines; and steamers ply regularly to the town of St Paul, the seat of government, situated immediately below the Falls of St Anthony on the Upper Mississippi,* and 219 miles within the borders of the territory.

Its geology, and the nature of its soils, have not as yet

* SEYMOUR'S *Travels in Minnesota, the New England of the West.*

been fully explored. The eastern portion of its surface appears, however, to consistof sandstones and magnesian limestones, forming sandy soils and naked prairies. The western part is an extensive chalk formation, forming prairies, in many places rich in pasture.

In connection with the lake-traffic which will necessarily arise out of the prosecution of the copper-mines of Michigan, the existence and rapid growth of these new States ought to be borne in mind. Their intercourse with Europe will be chiefly through the lakes, the Erie Canal, and the St Lawrence. The Erie Canal is already overloaded with traffic, and can never be enlarged to meet half the growing demands upon its means of transport. The St Lawrence, therefore, must become a great continental thoroughfare, a source of great wealth to Canada ; and if the Canadians would only exercise as much reasonable patience in the development of their natural resources as they have lately shown of political rashness, they may obtain from the United States any proper equivalent for a right of way through their canals and private waters.

Of all the wonders in rural economy with which this New World is pregnant, there is none with which the traveller becomes acquainted, in these lake-bounding States, which is more wonderful than the history of what is called the *hog-crop* of Ohio. This State, one of the most prosperous in the Union, and containing now a population of nearly two millions, is bounded on the north by Lake Erie, along which it has a coast line of upwards of 200 miles. Though this State possesses much wheat-land, and raises in some places large breadths of this grain, yet both its soil and its climate adapt it more generally to the growth of Indian corn. This latter grain, therefore, may be considered the staple of the State. The general fertility of the land, in comparison with that of the States of New York and Michigan—as they are at present cultivated—and the comparative productiveness of each crop in these States, will be seen by the following table, representing the average yield in each per imperial acre:—

	Ohio.	New York.	Michigan.
Wheat, .	$15\frac{1}{4}$ bush.	14 bush.	$10\frac{1}{5}$ bush.
Barley, .	24 ...	16 ...	...
Oats, .	$33\frac{3}{4}$ ...	26 ...	...
Buckwheat,	$20\frac{1}{4}$ ...	14 ...	...
Indian corn,	$41\frac{1}{4}$ ...	25 ...	...
Potatoes, .	69 ...	90 ...	...

In wheat it does not much exceed the State of New York, but it greatly surpasses it in the produce of Indian corn; while, from possessing a drier summer climate, it is inferior in the growth of potatoes.

Thus rich in Indian corn, an early question with the farmer was, how to dispose of his surplus growth. Until the famine years of 1847 and 1848, the quantity of this grain which was exported from the whole United States was exceedingly small. From 1820 to 1845, it averaged only about 60,000 quarters (half a million of bushels) a-year; but from 100,000 quarters in 1845, it rose to 230,000 in 1846, and to 2,125,000 quarters in 1847, after which it fell to half a million in 1848, and has since continued to decline.

A home outlet, therefore, was sought for, and the distilling of whisky and the fattening of hogs are the means of consumption which have been found most easily available, and most generally profitable. The extent to which the *packing business*, as the latter is called, is carried on, and the States in which the two conditions —the faculty of growing Indian corn and of producing abundant acorns—coexist most extensively, will be seen by the following table, which represents the number of hogs killed in each of the packing States of the west in 1846:—

Missouri, . . .	70,898
Tennessee, . . .	42,975
Kentucky, . . .	215,125
Illinois, . . .	68,120
Indiana, . . .	251,236
Ohio, ...	420,833
Other places, . . .	18,675
Total hogs killed, .	1,087,862

The hogs are allowed to run in the woods, and feed on the acorns, &c., till five or six weeks before killing-time, (8th or 10th November,) and are then turned into the Indian-corn fields, to fatten them and harden their flesh. They are usually from eleven to eighteen months old when they are killed; and the longer they have been in the corn-fields, the better is the pork.

The packing business of Ohio has been gradually concentrating itself in Cincinnati, where, in the winter of 1847 and 1848, about 420,000 hogs were sold, killed, and packed. The blood is collected in tanks, and with the hair, hoofs, and other offal, is sold to the prussiate of potash manufactories. The carcass is cured either into barrelled pork or into bacon and hams, and the grease rendered into lard of various qualities. Some establishments cure the hams, and after cutting up the rest of the carcass, steam it in large vats, under a pressure of seventy pounds to the square inch, and thus reduce the whole to a pulp, bones and all, and draw off the fat. The residue is either thrown away or is carted off for manure. One establishment disposes, in this way, of 30,000 hogs. Of the lard, the finest is exported—much of it to the Havannah—where it is used instead of butter. About thirty factories are engaged in the manufacture of lard oil and stearine, which is done by compressing the lard at a low temperature. The stearine is made into candles on the spot, of which 6000 pounds are manufactured every day, on an average of the whole year. The lard oil is partly sold as such, but in the Eastern States is used to adulterate spermaceti oil, and in France to lower the price of olive oil. It is said, that in this latter country, from 65 to 70 per cent of lard oil is often mixed with the olive oil without detection. The mixture is more apt to deposit stearine, however, than the pure oil, and such an appearance may lead to its detection.

The less pure lard, and the fat extracted from diseased animals and from the offal, is used in the manufacture of soap. Besides soft and fancy soaps, there are made at Cincinnati about 100,000 pounds of soap weekly, and of the fat employed about 80 per cent is pork grease. Lastly, the bristles give rise to a separate business, employing a hundred hands, and the hoofs are partly boiled down into glue.

The marketable products of the 420,000 hogs packed at Cincinnati may be thus summed up—

Pork (150,000 barrels,)	29,400,000	pounds.
Bacon, . .	21,000,000	...
Lard (No. 1,) .	13,800,000	...
Lard oil, . .	1,000,000	gallons,
Stearine candles, .	1,875,000	pounds.
Bar soap, . . .	5,200,000	...
Fancy and soft soaps,	7,300,000	...
Prussiate of potash, .	50,000	...

This is the yearly produce of a stock of about a million and a half of hogs in the State of Ohio. In the whole United States, the entire hog stock is estimated at upwards of 40,000,000. Hog-rearing must therefore be regarded as one of the most important branches of rural economy, and the hog-crop one to which yearly attention should be given.

This branch of economy in Cincinnati shows us how cities grow, how centres of united and simultaneous action become necessary to the most profitable development of rural industry, and how many arts arise out of—necessarily spring from, and are indispensable to—the profitable pursuit of the farmer's operations. No corn or hog grower in Ohio would venture to say that the interests of his class were unconnected with, or opposed to, those of the moneyed men and enterprising merchants and manufacturers of Cincinnati.

CHAPTER IX.

Case of American cleverness.—Fat cattle of Ohio.—Butcher in Buffalo. —Cause of the growth of the city of Buffalo.—Capital taken out by emigrants.—Influence of Europe on the progress of American cities.—Cause of the difference in progress of Canadian and New York cities.—Not a result of want of energy in the Upper Canadians. —Lake Erie.—Supposed periodical slow rise and fall in the level of the great lakes.—Evidence of such gradual changes of level.—Their relation to existing terraces, and ancient beaches.—Their supposed cause.—Water discharged by the Niagara River.—Hotel at the Falls. —Coloured waiters.—Geological Section at the Falls.—Published descriptions of the Falls.—Popular disappointment.—Wearing action of the water.—Varying amount of water discharged over the Falls.—Influence of the winds on Lake Erie.—Influence of the noise of the Falls on their impression upon the mind.—Railway to Lewistown. —View from the mountain ridge. Voyage on Lake Ontario.—Queenstown heights.—Profits of New York farming, by a New York farmer.—Knowledge and intelligence among these farmers.—City of Oswego.—Sackett's Harbour.—Railway to Canada.—Kingston in Upper Canada.—Character of the Upper Canadians.—Difference between a Canadian and a New York wife to a working man.—Difference in the character of the people in the States arising from the number of Germans among them.

Sept. 16*th.*—I began a previous chapter by an allusion to the use of the word *clever* in the United States; I introduce the present by an illustration of the "cleverness" of the people.

As we approached the end of our journey to Buffalo, a gentleman, to whom among many others I had been introduced at Syracuse, but whose name I did not know, accosted me in the railway carriage, and asked me to take up my quarters at his house, a couple of miles out of Buffalo. I excused myself from giving him trouble, on the plea that I intended to start again early in the morning for Niagara, and that it would be more convenient for me to go to the American Hotel. He then offered, while I waited for my luggage, to walk into the town to secure me a good room at the hotel. Accordingly, half-an-hour after, when I drove up to the hotel, I found him waiting, and comfortable quarters secured for me. In the morning, when I asked for my bill, I was told that everything was paid. I hesitated at first to receive this pecuniary obligation; but on reflecting that it was meant in kindness, I felt it would be unkind in me, in the absence of my unknown friend, to refuse it. I contented myself, therefore, with inquiring his name, and have pleasure in mentioning the circumstance here, as an instance of the proneness of our Transatlantic cousins to the virtue of hospitality. Notwithstanding the sour and exciting things said occasionally by bitter journalists, on both sides of the water, they will not, in our time at least, altogether forget that "blood is thicker than water."

The long and wide main street of Buffalo reminded me of the Trongate of Glasgow more than of any other street in Europe I recollect to have seen, though, of course, it is newer, and less finished in appearance. On the evening of my arrival I took a walk along it, to look at the many large and well-stored shops. Among others, I went into a butcher's store, in which the beef and lamb, to my eyes, seemed excellent. The prices of lamb and mutton were 3 to 6 cents, of beef 4 to 8, of pork 6¼, and of fowls, when full grown, 5 cents a pound. The import duty on Canadian beef is 20 per cent; so that fat cattle not reared at home are brought chiefly from Ohio, where, as I have said, the excess of Indian corn is used up in feeding stock. Besides the "hog crop" of south-western Ohio, the cattle crop of the eastern portions—which lie on the less fertile sandstones and non-calcareous clays of the Portage and Chemung groups of the New York geologists—is very large and valuable. The five counties of Ross, Pickaway, Franklin, Madison, and Fayette, send annually to market at least 35,000 head of fat cattle, worth £8 a-head.

After I had asked my questions of the butcher, and he in return had found out, by questioning me, first how many years, and then how many months, I had been in America, "Well, sir," says he, "we live in a great country here—we are a great people." I evaded what was meant as a question, and spoke pleasantly to his everyday ideas, by remarking that "I had certainly seen at Syracuse the very largest oxen I had ever beheld." So we parted very good friends, and he invited me to drop in and see him again.

It is unpleasant to a stranger to be always called upon to admire and praise what he sees in a foreign country; and it is a part of the perversity of human nature to withhold, upon urgent request, what, if unasked, would be freely and spontaneously given. But highly to esteem, and value, and prefer one's native or adopted country, is a virtue which is to be commended and encouraged. It is the basis of individual mental contentment, and of that general patriotism which has in so many countries led to great and noble actions, and which has always ranked the first among political virtues. If a man does not think the country he lives in the best in the world, he had better leave it. But this does not justify or excuse either unfounded arrogance or self-esteem in a people, or the tendency to brag and swagger which one does occasionally see among individuals in the United States.

Buffalo, as I have already remarked, is a very thriving town, and the causes of its success are very intelligible, though not always clearly seen or fairly put by writers on the rise and growth of American towns.

Situated at the end of the Erie Canal, with a large American population and the highway to Europe behind, and with boundless tracts of new land before it beyond the lakes, Buffalo, like New York, has risen from the mere force of circumstances. Every emigrant, and every package of goods, that passes through New York, Albany, and Buffalo, imparts some gain to each place in the transit. And as the city of New York increased with the western population of the state, so Buffalo has increased, and will increase, with the number of persons in the new north-western States, whose way to and from the principal markets is by the Erie Canal.

So also the new States rise in numbers and wealth. The poorest of the Irish immigrants who land at New York, Boston, Philadelphia, or New Orleans, bring with them some money—the greater number enough to pay the travelling expenses of their families, to buy a piece of land, and to maintain them for a year. The fare alone from New York to Chicago, in Wisconsin, is fifteen dollars a-head, which is about £10 for a man and his wife and two children. The English and Scotch and German emigrants appear to be better and more thoughtfully provided for than the Irish; but Pat's ragged coat, as the captains of steamers know well, often conceals more gold than the decenter garments of the emigrants from other countries.

Taking rich and poor together, it is a very moderate assumption that the emigrants, on an average, carry out £10 a-head, which, for the 200,000 who land at New York alone, makes the sum of £2,000,000 sterling added at once to the money capital of the districts through which they pass, and in which they settle. Then a single year's labour of this 200,000, in agricultural operations upon new land, must add at least £5 a-head, or another

£1,000,000 to the capital of the new States; while the increased consumption of imported articles, by the added population, augments the Federal revenue which is derived from the duties levied upon imports.

It is Europe, not America, therefore, that is the cause of the rapid growth of the United States—European capital, European hands, and European energy. If all the native-born Americans—not being the sons or grandsons of Europeans—were to sit down and fold their hands, and go to sleep, the progress of the country would scarcely be a whit less rapid, so long as peace between America and Europe is maintained.

It is thoughtless in travellers to contrast the towns of Buffalo and Rochester and Oswego, on the New York side of the lakes, with Colburn, at the mouth of the Welland Canal, on the Canadian side of Lake Erie—or with Toronto and Kingston, on the opposite coasts of Lake Ontario; and to draw comparisons unfavourable to Canadian energy and enterprise, from the relative prosperity of these several places. There is quite as much energy in the blood of Upper Canada as there is in the British and German blood of western New York. But the local position of these towns of Upper Canada, and the condition of the inner country, forbids their becoming, for many years, equal in size or in wealth to the towns I have named. Suppose Colburn, like Buffalo—being at the end of canal navigation—had as large and growing a population behind it, and as extensive and valuable western territories before it, and that the highway from Europe lay through it instead of through Buffalo, then Colburn would have rivalled or exceeded Buffalo, even at this early period of their several histories. But this slow town of Colburn, as many have thought and called it, has nevertheless a great future before it. The natural outlet of this western region is by the St Lawrence. The Erie Canal is already unable to accommodate its traffic, and as this increases with the growth of the North-western States, more and more of it must proceed by the Canadian canals and waters, and drop its fertilising contributions as it passes through the country. With the settlement of the interior, also, and the increase of means of inter-communication, Toronto, as the natural course of the cross country traffic from Lake Huron; and Kingston, from its situation at the head of the St Lawrence, will both become seats of commercial wealth, and towns of political importance. I am sure that, if my Canadian fellow-subjects will be content to wait patiently for the natural course of events, which no Government or energy can precipitate, but which domestic disturbances will much retard—most seriously, perhaps, by their effect upon European opinion as to the desirableness of the Canadas as a place of settlement—they will soon see every reasonable expectation fulfilled. Even now, instead of granting that they are justified in looking either with envy or discontent at the growth of other places, I can only see reason to wonder that, in their geographical position, and with their political fretfulness, they have of late years increased so wondrously fast.

The morning was fine as I left Buffalo in the railway, which skirted the foot of the lake, and for a great part of the way ran along the banks of the Niagara River. The waters of Lake Erie, rippled by a light wind, were of a beautiful blue, and the numerous vessels to be seen on it made the view to me very interesting. To see on the far horizon large ships crowded with canvass, looming through the thin haze, and blue misty hills in the distance, such as one observes with eagerness when making the land after a long voyage across the ocean, realised to my eye that this fresh-water lake is, in reality, a great inland sea.

Before leaving this lake, I may advert to one important physical circumstance, in regard to the level of its waters, to which my attention has been drawn. It is the result of long observation that the surface of this, and, I believe, of the other lakes also, is subject to gradual, and, as some believe, periodical, but certainly very considerable alterations of level.

This is proved by the observations of those who reside on their shores. The water on sand-banks becomes shallower or deeper. Mills at the mouths of streams are rendered permanently useless by the rising level of the lake into which the streams descend. Former roads along the lake, as that immediately beyond Buffalo, have been overflowed, and rendered impassable. Old beaches, covered with trees and cliffs, are seen far inland, showing the greater height to which the waters formerly attained; while others, which men remember to have been at a distance from the lake, have again been reached, and are in progress of being undermined.

The height and periods of this rise and fall are both uncertain. In 1838, Lake Erie reached the highest elevation it has attained during the present century, and since that time it has been gradually receding. In 1788 or 1790, it was higher than in 1838, after which time it receded probably for many years, and then began again to rise. On the shores of Michigan, the rise was estimated at 5¼ feet between 1819 and 1838; and in another place, a resident of twenty-three years on the spot observed that, though it was highest in 1838, it was, in 1840, still 4 feet higher than when he settled there in 1817.

Facts of this kind have long drawn the attention of the inhabitants along the lakes. They are to them a source of anxiety, and, where low flat lands stretch along either shore of the lake, of alternate gain and loss. They are also interesting to the geologist, in connection with the ancient terraces, and the frequent more or less distant sea-beaches which skirt the lake-shores at various places. How high the lake may rise, when it next begins to increase, past experience does not enable us to judge. As we are ignorant of the cause, we cannot say to what level a rise is possible in existing circumstances; we cannot, therefore, reason as to the cause or antiquity of the ancient lake-beaches, especially those which are not greatly elevated above the present waters, or draw safe conclusions as to the permanent change of level which the lakes may now be presumed to have undergone. Hence these oscillations in the lake-levels have been subjects of inquiry and discussion both by Mr Hall, one of the geologists for the State of New York, and by Mr Higgins, of the geological survey of Michigan.

Variations in the fall of snow and rain in the lake country, and differences in the amount of evaporation, suggest themselves as the simplest causes of the phenomena. But such causes—unless, in this region, they act in obedience to some steady alternating law—will not explain the specialties of the case. The rise and fall of the lake-levels are so gradual, and continue to augment for so long a period, that a steady and increasing augmentation of the water poured into the lakes must go on while the level is rising, and a similar gradual and long-continued diminution while it is falling. Meteorological observations have not yet shown that such augmentations as these of the fall of rain and snow, or of lake evaporation, do take place in any part of the world.

The quantity of water which escapes from the lake by its natural outlet, the Niagara River, is an important fact in this discussion. The Falls of Niagara, during the high-water of summer, allow 20,000,000 of cubic feet to fall over them*—a discharge which, taking the area of Lake Erie at 10,000 miles,† would lower the

* This is Mr Barrett's, the latest and best determination.

† Dr Houghton estimates it at 9600 square miles.

level of the lake one foot in ten days; or, more correctly, if Niagara were dammed up, the level of the whole lake would be raised two feet in ten days—one foot by the usual supply of water poured into the lake from above, and another by the water prevented from escaping. Any sudden increase in the fall of rain or snow, therefore, would soon run off, and leave the lake at its usual level.

If we consider the case of Lake Erie by itself, and compare its area of 10,000 square miles with that of all the upper lakes united, which cover 77,000 square miles, and suppose these upper lakes to be raised very high by one extraordinary fall of rain or snow, or by a great diminution in the evaporation of one short cold summer, then it is possible that the discharge from these upper lakes of this one unusual amount of water might, by the nature of the outlet into Lake Erie, be so regulated as to continue augmenting for a series of years, and again, as it lessens, to continue diminishing for another series. But this possible explanation seems to fail, when it is recollected that Lake Michigan itself, one of these upper lakes, exhibits similar oscillations of level. The source of the increased or diminished supply must, therefore, be sought for in Lake Superior, if this be considered a probable cause. But, unfortunately, the remoteness and hitherto generally wilderness state of the shores of this lake have prevented any observations being made, by which light could be thrown on this interesting point.

Part of the country through which the railway conducted us on our way to Niagara was still uncleared or unstumped, and sprinkled with log-huts and apparently poor settlers, surrounded by indifferent crops of Indian corn, on soils evidently better adapted for wheat. We cross again, on this route, the belt of flat wheat-land, belonging to the Onondago salt and Niagara limestone groups, which, as I formerly stated, stretches beyond the Niagara River far into Canada. As seen here, it is a clayey region, on which the system of thorough-drainage is destined hereafter to produce most beneficial results.

I reached the Falls of Niagara, on the American side, at a quarter past ten, in time to hear service well performed in a new, nicely-finished, though small Episcopal church. This village of Niagara consists chiefly of hotels and churches; and the running of a morning and evening train to Buffalo is considered indispensable to the success of at least one of these sets of establishments.

At dinner at the Cataract Hotel, we had a large party of about a hundred and twenty, though not half as many as the room was fitted to accommodate. This universal dining in public, in the United States, of all sexes and ages, is one source of the forward boldness of so many of the young people. And although the mingling of all classes at these tables teaches the use of silver forks to persons who would never meet with such things at home, yet it roughens the general tone of speech and manners of all, and makes those who really know better fall into customs they would at home be the first to reprove.

I may remark, however, that perhaps too much is said by travellers of the solecisms of guests at the American tables. I doubt very much if a similarly indiscriminate assemblage of persons of all classes at an English table would, on the whole, behave so well. Besides, the custom in the American hotels of loading the table at breakfast and dinner with a countless number of small dishes, not half of which are furnished with knives and forks, or spoons to lift their contents, leaves the majority of the guests no other resource, than either to be helped with their own, or probably to deny themselves what the dishes contain altogether, and leave them to less scrupulous neighbours.

What amused me most at this hotel was the excellent discipline maintained by their *chef* among the eighteen black waiters who attended the table. In carrying out the first course, they all started at a signal, and marched *en militaire* in double file, each bearing his dish, and presently returned in the same order with the second course, opening into Indian file as they reached the head of the table; and when each had reached his station, depositing the whole at the same instant on a signal from the head-waiter, who was also dark-coloured. The peculiar proud swagger with which all this was done, the air of the men as they strutted along, and the evident "Isn't that well done?" which each of them looked as he lifted his cover, were most amusing to me, who had not yet had much opportunity of studying the peculiarities of the free coloured people in the northern States. My own sympathies have always followed this unhappy race of people, whether in slavery or in freedom, and I have usually found them civil and obliging. They are often, however, very conceited; and can be very saucy, as white servants in English hotels not unfrequently are. But they are in general very quiet and civil, and have a peculiar knack at waiting. Of absolute rudeness among this class of people, the only instance I met with was in the Irving Hotel in New York, where black servants are employed, and where, on the occasion of my visit, one peculiarly black and impertinent sheep had certainly found a place among the flock.

In the afternoon, I went down to the Falls. I crossed over to the Canadian side, and spent several hours on the banks which overlook them. I afterwards walked to the suspension bridge a couple of miles below, which is itself a nervous thing to walk along, and from which

the view of the Falls, and of the ravine, is striking and beautiful. The section of the strata, as seen at this place, is as follows:—

Limestone, . Shale, . .	.	Niagara group.
Limestone, .	.	Clinton group.
Sandstone and thin clay marls, chiefly red,	.	Medina sandstone.

This section is now well known, as well as the influence of the Niagara shale, in hastening the working back of the Great Falls. It illustrates, however, what I have had occasion to say in reference to the soils and geology of western New York. The numerous layers of red clay marl, among the red rocks of the underlying Medina sandstone, are in conformity with the economically important observation, in reference to the agricultural value of this group of rocks, to which I adverted in the preceding chapter—that the poorer Medina sandrock of the eastern counties of New York becomes more mixed with clay towards the west. Hence the rich soils to which it gives rise below the mouth of the Niagara River, and along the south-western borders of Lake Ontario, where it forms the surface of the country.

Above the Niagara limestone, rest the Onondaga salt rocks and their *debris;* and though these are spread over the surface of the country in the neighbourhood of the village of Niagara, they are not seen in the section of the ravine as it appears from the bridge, nor on the immediate banks of the river.

I attempt no description of the Falls. The first peep I had of them showed me how very little all I had read of them had impressed me with anything like a definite idea of the peculiar features of this great descent of water, or of what I was entitled to expect when I came to look upon it. I infer, from this, that they cannot be adequately and graphically described—or, at least, that I should fail were I to attempt to do so. I have seen many water-falls in many countries, and I venture only to remark, in regard to this one, that I think it is a piece of great presumption in the common class of tourists to talk as one usually hears them do, of their having been disappointed—as if some great showman had got up the thing for their amusement, and had not put gunpowder enough into the crackers sufficiently to astonish their weak minds.

On the Canadian side of the Falls, a high bluff of red, probably drifted clay, rests above the Niagara limestone, forming an upland above the narrow fringe which separates it from the waters of the river above the Falls. Below the falls, this bluff retires to a considerable distance from the river, and the carriage-road to the suspension bridge runs along the surface of the nearly naked rock. When walking leisurely here, two things agree in forcing the same thought upon the imagination. Where it is completely uncovered, the whole upper surface of the limestone rock, on which we travel, exhibits evidence of the wearing action of the water. It has the same hollowed and irregular appearance as the surface above the falls, over which the water is now pouring. Over this, therefore, the river must formerly have run, before it ate out the deep ravine below. And, again, the retiring of the bluffs shows that it then, as we should suppose, had occupied a wider bed, and, as it now does above the Falls, had undermined the cliffs of clay, and bent its course now more to the one side, and now more to the other, as circumstances might direct. One reflects on such things, and in his closet makes cool calculations of the lapse of time necessary to accomplish all this. But the greatness of the lapse is felt when we see before us the protracted effect, and the still living and acting cause. The foam of the cataract becomes, to the imagination, the hoary hair of thousands of years, and its perpetual rainbow a halo round the head of the sleepless spirit which has seen these changes and survived them all.

The quantity of water which falls over these rocks has been variously estimated. The most trustworthy, perhaps, is that of Mr Barrett, already quoted, which was deduced from three different observations, made at Black Rock during the high water of 1838 and 1839. This estimate makes it amount to nineteen and a half million cubic feet per minute. It varies very much, however, with the height of the water and the direction of the wind. When a strong wind blows from the west, the water at the east end of Lake Erie will rise several feet in a few hours; and so much more water is driven down the Niagara River at such times that the river in the ravine below the Falls, though so rapid, "frequently rises 15 or 20 feet during a westerly wind." A wind from the east produces a contrary effect, lowering the water at the east end of Lake Erie, and lessening the quantity of water which passes over the Falls.

September 17.—The forenoon of this day I spent chiefly on Goat Island, wandering about the Falls on the American side. There is nothing to be seen from this side which can compare with the quiet and graceful beauty of the American Fall, as seen from the Canadian side. The perfection with which the folds of that broad, living, spotless stream are draped together cannot be imagined; and though there are many beauties among which days are too little to spend on Goat Island, the quieter spots were to me the most attractive. In truth, the reason why the disappointed people talk of the thing growing upon them is, that they must become so familiarised with the noise and roar as to be able to abstract these altogether from the scene, before their eyes and hearts can come into independent contact with its true attractions. This abstraction, to the generality of minds, is a difficult task; but, once made, then the sound of the cataract comes in like far off music—not in the foreground, deafening and annoying, but soothing and lulling, as it were, in the distance, and thus administering to the enjoyment it had formerly intruded upon and broken up.

The fresh-water shells which occur in the deep bed of mixed slaty gravel and red clay drift, which covers the limestone rocks at the edge of the waterfall upon Goat Island, are now well known. The minute, almost microscopic species, I found very abundant in the clay. Besides the shells usually collected, I picked up a fragment of a fresh-water crustacean.

After dinner, I left Niagara, on my way to Lewiston —where the river escapes from the ravine and the high lands—to take the steamer down Lake Ontario. This railway is a very indifferent affair, and Lewiston a long straggling skeleton of a place; but the last mile's ride along the edge of the ridge, and the view it affords, is worth going twice as far, and by a still rougher road, to enjoy. Sheer down we looked from a high escarpment, upon the broad flat forest lands, stretching many miles back from the lake, and along its shores farther than the eye could reach. Here and there only, in all this distance, a clearing appeared; while right before us lay the endless lake, and its occasionally bolder shores beyond, with now and then a straggling sail or a distant steamer's smoke, and all mellowed and blended together by a four o'clock sun. The whole prospect perhaps struck me the more that it came upon me quite unexpectedly; but I regretted much that previous arrangements did not admit of my spending a day in wandering over these high lands—scarcely allowed me even a moment to master what was before my eyes, as the steam was up, and bags and portmanteaus demanded the attention of the careful traveller.

The mountain-ridge, as it is called, formed by the outcrop of the Niagara limestone, has been long known to the inhabitants of western New York. When, some two centuries hence, all the low plain beneath it shall be cleared, and drained, and cultivated, and smiling villages, and cheerful homesteads, and scattered flocks and herds, overspread its surface, and the blue smoke dies away from many chimneys as the Sabbath-bell draws the gathering people toward the frequent house of worship—how many, in those days, for broad pictures of wide natural beauty, intense with countless little episodes of still life, will frequent this mountain-ridge, when the noise of the cataract has wearied them, and they wish again to calm and compose their spirits, worn out by its ever-fretting impatience!

The escarpment which forms this ridge is bolder above the village of Lewiston than it is in some other parts of its course along the southern shores of Lake Ontario. The following section (taken from Mr Hall) gives an idea of the physical and geological nature of the ridge itself, and of the flat country and its soils which lie below—

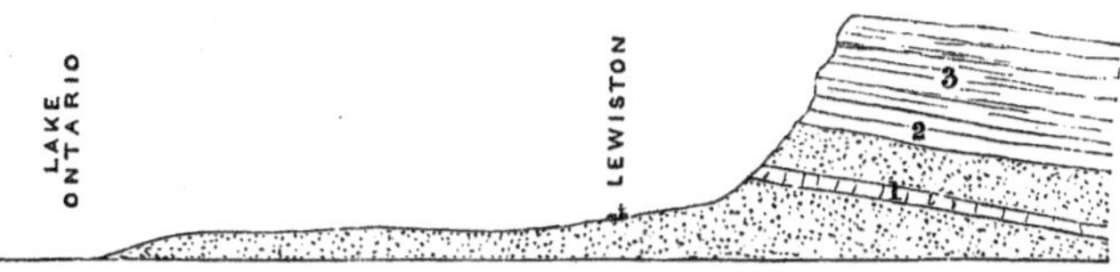

Here the section No. 1 is the Medina sandstone, consisting chiefly of red sandstones and shivery clay marls. No. 2 is the Clinton group, of no great thickness or consequence; and No. 3 the Niagara shale, surmounted by the Niagara limestone. The long flat edging of the lake consists of the red soils of the Medina rocks, and ought to be very productive. In many places, from its level character, it is marshy or full of water; and in many others arterial drainage on an extensive scale will probably have to be introduced, before the capabilities of the country can be at all fully developed.

The escarpment of the Niagara limestone is not everywhere, as I have already remarked, so high nor so abrupt as it is represented in the above section; but, with occasional breaks, it may be visited along a great part of western New York, with the certainty of commanding from its summit an extensive view of the flat country below, and of the wide blue lake beyond. It extends also, towards the north, round the western end of Lake Ontario, and then eastward for many miles, forming an escarpment far behind Toronto, carrying with it into Upper Canada a wheat-region, not unlike that of western New York.

As we steamed from Lewiston through the mouth of the Niagara River, and entered the lake, Queenstown, on the Canadian side, appeared to us on the water to be at least as flourishing as Lewiston, which we had just surveyed by land. The heights above it, on which opposing forces of the same blood, with equal gallantry, fought the battle of Queenstown, and where the well-known pillar commemorates the fall of the brave Sir Isaac Brock, are as high as the ridge above Lewiston, of which I have spoken, and promise to the lovers of the picturesque as wide, and if it bear the eye to the western extremity of Lake Ontario, a wider, and perhaps a still more beautiful view of mixed land and water, high-land, forest, and cultivated fields.

On board the steamer, I had a concluding conversation on the profits of farming in western New York, with a practical farmer from Syracuse. "The results of my personal experience are," he said, "that money is not to be made by farming in this State. If a farmer hire *two* men, and work with them, and keep them at their work, he may maintain his family, and clear 8 per cent upon the value of his farm. But if he farm more largely, as a gentleman farmer, leaving the management to an overseer, he will not make more than perhaps 2 or 3 per cent. Farming is much less profitable in my county of Onondaga, during the last five years, than it used to be. Exhaustion has diminished the produce of wheat, formerly the great staple of the country. When the wheat fell off, barley, which at first yielded 50 or 60 bushels, was raised year after year, till the land fell away from this also, and became full of weeds. It still grows 50 bushels of Indian corn, and this is the best crop we now get—but it must be manured. Much is now laid down to grass to be recruited; but those who are anxious to make money are turning their hands to something else, and either selling or letting their farms. A farm in a good situation can be let to pay 5 per cent; but as 7 per cent is easy to be got for money, few persons care to continue the owners of farms which they cannot cultivate themselves, and can only let to yield a return like this."

Such was my new friend's opinion of agriculture in the empire State; and I have since met with many who agreed with him in all essential points. No interest in national importance can ever, in this New World, compete with the agricultural; and yet, after the first two or three generations, the most energetic and aspiring sons of the first pioneers forsake the scene of their fathers' labours, and betake themselves either to other pursuits or to new regions. A certain numerical strength, permanent competency, and abiding position in long-known localities, will always in every country remain with the owners of the land; but in the United States, as elsewhere, the energy and activity, and intellectual influence upon social and political progress, will be mainly possessed by the alert and enterprising brood who yearly hive off from the old lichen-covered stationary stock.

I shall have occasion, hereafter, to speak of the prospects of agriculture, and of the progress of rural improvement in general, in the State of New York. I will here only add, in justice to the New York farmers, that, in my journey to Buffalo, I was struck with the very general familiarity which seemed to prevail among practical men as to the geological character of their country, and the relation which the geological details had to the agricultural qualities of their farms. The efforts of the State Government, in distributing numerous volumes of their Natural History Survey, has no doubt aided much in diffusing this knowledge, so rare in an agricultural community, and yet so creditable to their intellectual position. In the county of Surrey only, and along the borders of the chalk and green-sand country of England—where geological differences affect the soils in so marked a manner, and within such short distances—have I met among practical farmers, with so clear an idea of local geological relations, and of their connection with rural labours and profits.

As the boat paddled out into the lake, the water roughened a little, and many of the passengers became indisposed. I had a comfortable state-room; but the meals, which were included in the fare, were uncomfortable and crowded. It was little else than a rush and a scramble, where the slowest and weakest got neither place nor pudding. At 11 P. M. we reached Rochester, a distance of 80 miles, and after a short delay steamed on to Oswego, 65 miles farther, where we arrived at 6 in the morning, and were at first allowed time only for a short stroll, but finally were detained till half-past 10, in waiting the arrival of the train from Syracuse.

The flower-mills of Oswego are, I believe, the staple source of its prosperity. It is a thriving town, and, being at the termination of the canal and of the railroad—both of which connect it directly with the Atlantic, while the lake lays open all Canada and the west—its commerce and importance are likely to augment. In anticipation of this, it has already been incorporated into a city. My readers must not interpret this as indicating a very rapid or high degree of prosperity, since the language of this State, as I have already explained, recognises nothing between a village and a city.

Through this port, a large part of Upper Canada is supplied with salt from the salines of Syracuse; and along the wharfs lay numerous little mountains of salt-casks, waiting to be shipped.

From Oswego to Sackett's Harbour was a run of forty-five miles, along a shore which is still very much wooded. Now and then, in the midst of the forest, a few burnings became visible, showing the work of clearing to be still in progress.

Beyond Sackett's Harbour, on its north-eastern side, is laid out the railway now in progress, which is to connect Kingston in Upper Canada, across the Ten Thousand Isles, with Rome in the State of New York, and thence by the existing line with Albany. It will be a great boon to Canada when it is finished, and a matter of much moment also to the city of New York.

The village of Sackett's Harbour shows nothing to arrest attention, beyond its hotel, and some signs of increase in size. After stopping about an hour we started again for Kingston, which is forty miles across the lake, making in all a distance, by this route, of two hundred and thirty miles from Lewiston to Kingston.

The harbour (Sackett's) and the islands about its mouth, and among which we steamed, made this part of the voyage very pleasant. We passed the main channel of the St Lawrence, which is about a mile in width, bore up N. by E. at Simcoe Island, on which the lighthouse stands, and then sailed a straight course of many miles for Kingston. Long before the eye could make out the houses of the city, the towers of a lofty building raised themselves like a beacon, far above every other object in massive proportions and in height. As we neared the coast this proved to be the Roman Catholic cathedral, occupying the highest ground in the city. In many other parts of North America, as well as here, I have found the Romish churches ambitiously securing the most prominent and imposing positions. They are all selected with an eye to the future.

At 6 P.M. we landed on the pier. I almost felt myself at home again as I set my foot on shore in sight of the British flag; and the kind welcome of a Kingston family added double pleasure to the agreeable week I subsequently spent in this place.

In manners and in sympathies, a sensible difference still prevails between Upper Canada and western New York. Notwithstanding the proximity of the two countries, and the increasing intercourse between them, this will probably long continue to be the case.

Part of the difference which is felt, in crossing from either side, may be in idea only, and connected with one's political prejudices, republican or monarchical; yet sensible differences, both in men and women, exist nevertheless. One feels the *de trop*—the tendency to exaggerate—among the men on the one side, obtruding itself sometimes offensively, especially in the newer States of the Union, and among the newer people. An opposite tendency, and not unfrequently symptoms of discontent lurking at the corners of the mouth, are met with along the Canadian border, so often as to arrest attention to the circumstance. But the Upper Canadians have in themselves, and in their country, the materials of a first-rate people, if their eager spirit, anxious too speedily to excel, would permit them to proceed steadily on their way.

The Upper Canadian women have their character too.

"I'll go over to Canada for a wife when I marry," said a young south-shore farmer to his friend. "When I come home at night she'll have a nice blazing fire on, and a clean kitchen, and a comfortable supper for me; but if I marry a New Yorker, it'll be, when I come home, 'John, go down to the well for some water, to make the tea;' or, 'John, go and bring some logs to put on the fire, to boil the kettle.' No, no; a Canadian woman's the wife for me."

One circumstance which will materially modify the population on the opposite shores is the large number of Germans who have settled in the States, while the population of Upper Canada is almost wholly British. This, I think, promises a more active future to Canada than the population of New York would give her.

I have already drawn attention to some of the places in the valley of the Mohawk, where Germans especially abound. But all along the western region they are numerous. In Buffalo, the German correspondence is so extensive that a separate bureau, as I was informed, is established at the post-office for German letters; and in the far north-western State of Minnesota—the New England of the West—the annual message of Governor Ramsay to the Legislature for 1850 was printed in German as well as English, which shows how many of that tongue are already numbered among its adventurous inhabitants.

CHAPTER X.

Kingston.—Soils of its neighbourhood.—Importance of agriculture in Canada.—Show of the Upper Canada Agricultural Society.—Pork-raising in the provinces.—Adaptation of breeds to local circumstances.—Implements in the show-yard.—Infancy of root-culture in the province.—Alleged difficulty in the turnip-culture.—Rocky Mountain beans.—Canadian coffee.—British sympathy with colonial grievances.—Alleged pusillanimity of the Governor-general.—Farming in Home district.—Wheat the surest crop in Canada West.—Excellence of the winter wheat.—Best wheat-belt round Lake Ontario.—Total produce of Canada West, and average yield per acre.—Large consumption of oats.—Less productiveness of the wheat-crop than in former times.—Cause of this.—Social position of the farming class in Upper Canada.—Means of improvement now in progress.—United Empire Loyalists.—Limited capital of the farming proprietors.—Condition of the grants made to the United Empire Loyalists.—Renting of land, and farming on shares.—Indian-corn whisky, and malt.—Extensive manufacture of such whisky at Cincinnati in Ohio.—Use of Indian corn in the Canadian distilleries, and of mixed rye and pease.—Whisky from pease.—Prospects of Kingston.—The Ten Thousand Isles of the River St Lawrence.—Descending the rapids.—The Sault St Louis.—Nature of this rapid, and of the descent.—Approach to Montreal.—Metamorphic limestone rich in phosphate of lime.—Agricultural value of this rock, and of the mineral phosphate as an article of export.—Deposits of mineral phosphate in the State of New York.—Origin or source of this mineral phosphate, and of garnet, graphite, and other minerals found in crystalline limestones.—Graphite and phosphate of lime in an altered rock, evidences of the former presence of organised bodies.—The crystalline limestone interstratified with gneiss.—Singular contortions exhibited by the limestone.—Dr Emmon's explanation of the cause.

Sept. 19.—Kingston, on my arrival, partook of two different forms of excitement—one in common with the whole province, arising out of the political differences and recent burnings at Montreal; the other—peculiarly its own—connected with the show of the Agricultural Society of Upper Canada, which was to be held in its suburbs during the course of the week. The interest taken by all parties in the political question was great; and the comments on the procedure of the Government and the Governor-general, numerous and free. The influence of all this was not insensible upon the affairs of the agricultural meeting, at which, until almost the last hour, his Excellency's presence was expected. On the whole, however, the meeting went off very well, and very creditably to the agriculture of the province—though with the introduction into the after-dinner speeches of more politics than would have been permitted by an experienced chairman on such occasions at home.

The neighbourhood of Kingston is an inferior agricultural district. A thin, generally light soil, rests on the Trenton limestone—a solid deposit of dark-blue fossiliferous rock, which here skirts the northern shores of the lake, and extends inland to a considerable distance. The richest lands of this division of Canada lie towards the north and west, a portion of the province which I regretted that my previous arrangements did not permit me to visit.

In Canada, every one is satisfied of the paramount importance of the agricultural interest; a very general desire exists, therefore, to advance it by every reasonable or available means. The superior class of settlers, of whom so many are scattered over Upper Canada, will greatly facilitate the adoption of such means of improvement as are usually employed, or are easily available by agricultural societies.

The Agricultural Society of Upper Canada had been in existence only three years, and the excited state of political parties had retarded that general union, even upon the question of rural improvement, which, as men's minds sober down, must eventually take place. Still I was agreeably surprised, both at the extent of the preparations I saw making on my arrival, and with the appearance of the town and of the show-yard on the day of the exhibition. The latter was not so extensive nor so crowded as that of Syracuse, but much more numerously attended by well-dressed and well-behaved people, and rendered attractive by a greater quantity of excellent stock and implements than I had at all anticipated.

The best of the stock was brought from the western part of the province. Among them were superior Short-horns, a few Devons, and some Ayrshires—all of pure blood. The greater number, however, were crosses, which, as in the States, are here called *grades*. The Leicester sheep were very fine, and the prize pigs—chiefly Berkshires—excellent.

The pig husbandry in Canada and in the province of New Brunswick, to be conducted economically, requires to be somewhat modified in comparison with the method adopted in Ohio, and the other large hog-growing and Indian-corn producing States. Of the vast number slaughtered at Cincinnati after harvest, the ages vary from a minimum of eleven to a maximum of nineteen months. They are generally kept over one winter, and *packed* before the next commences. In the provinces, the first difficulty which the settler has to overcome is that of laying in a sufficient stock of food for the long months of winter; and although the introduction of a better husbandry will by-and-by greatly lessen this difficulty, yet at present it is a main object with the farmer to get the winter over at as little cost of food as possible. The aim in regard to pigs is, therefore, to obtain a breed which shall litter in April, and can be fed to produce a barrel of pork (196 lb.) in November or December of the same year, and thus to save all winter keep, except for the breeders. As the lumber trade retires farther back, and becomes less extensive, the large and fat pork which was in demand for the lumberers becomes unsaleable, and a new form of the article—such as a civilised community are likely permanently to consume—is necessary to be produced.

Considerations of this kind render it necessary to look at stock in different countries with a differently instructed eye; and the opinions of a committee in offering, and of a judge in awarding, prizes, must be determined, not so much by the abstract excellence of this or that animal or breed, as by its special adaptation to local circumstances, and to the purposes for which it is reared.

Among the implements, which considerably exceeded in number and variety what is often to be seen at the shows of the Highland and Agricultural Society of Scotland, there were many excellent ploughs, harrows, cultivators, &c., manufactured in the province. Straw and corn-stalk cutters, and corn-shellers, were in considerable numbers; but an English visitor would be struck by the want of drill-machines, and root-cutters, and grain-crushers, now so abundant at our British shows. A few of the former, for drilling wheat, were the only implements of the kind which were to be seen. The roots exhibited—turnips, carrots, beet, mangold-wurzel, &c.—were all large and fine, showing the aptitude of the climate and soil to this culture. But here, as elsewhere in North America, the root-culture is still in its infancy. A rich virgin soil, producing crops for many years almost spontaneously, gave no stimulus to the preparation and preservation of manure among the first settlers; while the ready sale for wheat, and the difficulty of procuring hay for large winter stock, have hitherto prevented the attention of the existing farmers from being turned to the rearing of cattle.

Besides the fly—which, as with us, commits here also

great ravages upon the young turnip—drought is said to prevail in this province more frequently than at home, at the time for sowing the seed. There is greater difficulty also, it is alleged, in keeping the roots during winter than with us. Such difficulties, however, appear more formidable at first; experience generally shows how they are to be overcome. The last-mentioned difficulty has often been put to me in the province of New Brunswick, chiefly by persons new to the turnip. Yet even there I have met many practical farmers, who had fairly entered upon the culture, and had experienced its benefits in the winter-feeding of their stock, to whom the preservation of their turnips, in cellars of a properly adjusted temperature, had presented no difficulty whatever.

Among the horticultural productions were two which were new to me. One was the Rocky Mountain bean, which had pods from 12 to 18 inches in length, growing in pairs, and about the thickness of a common French bean. The seed has the appearance of a small kidney-bean. The other was labelled *Canadian Coffee.* It is a species of pea growing in a small inflated pod. It has the flavour of a pea, with a bitterish after-taste, and when roasted, has much of the odour and taste of coffee. It grew readily and ripened in the neighbourhood of Toronto, and may possibly come to be economically important.

On the whole, as I have said, this Kingston show was very creditable to the province of Upper Canada. The thousands of people who came to it, the stock and implements exhibited, the respectable appearance, the orderly behaviour, the comfortable looks and cheerful faces of both male and female, spoke for a state of things at least not very unflourishing. The British blood is purer in Upper Canada than in the State of New York, where, as I have already remarked, Dutch and

German settlers occupy large portions of the territory, and crowd into the towns; but in both there is enough of its influence and energy seen everywhere to make a home-born man proud of his country and his people. Faces, persons, dispositions—all look like home over again. The most pushing and impatient of the colonial-born little imagine how very much they resemble the tens of thousands of men at home who restlessly gnaw the bit of restraint—by which order can alone be secured, and leisure obtained for that cautious and steady progress by which advances, economical and political, which all consider desirable, may be safely made and successively rendered secure. I venture to say for John Bull, and Sandy too, that there is not a single British colony to which it does not delight them to hear that their brothers and cousins are going, or, having gone, to learn that they are prospering in it. Nor is there not a single real grievance with which any of these colonies may be afflicted that does not meet with their ready sympathy, and, whatever party may be in power, their ready co-operation, by all lawful means, to secure redress.

I introduce these observations, as the reader will readily understand, in reference to topics which I often heard discussed during my stay in Canada. At Kingston, the presence of his Excellency the Governor was looked for by many; and the number of guests at the Society's dinner was probably very much fewer in consequence of his absence. Various comments, of course, were made, according to the feelings and wishes of parties, upon this absence. His friends were disappointed, as they were ready to support him. His enemies called it pusillanimity, and used many other hard words. For my own part, I knew too little, at the time, of the temper of the people, to permit me to form an opinion as to the propriety of his coming or staying away. I was from the first, however, inclined to regard the Governor's movements as impelled chiefly by a prudent caution, and all I afterwards learned or saw tended to confirm and strengthen this first impression. What matters the personal bravery of the Governor of a colony, compared with his personal prudence? To satisfy his partisans, or to disprove the allegations of his opponents, is he to place himself in circumstances where disturbance may occur, the public peace be endangered, lives possibly sacrificed, and feelings of bitterness awakened which he may never be able to allay? The Annexation party in Canada would justly have blamed his rashness, his imprudence, and his unfitness to preside over a constitutional government, had they succeeded, by their taunts, in compelling him into any equivocal position.

The farming about Toronto—in the Home district, which I was unable to visit—is said to be better as to implements, stock, and handiwork, than in any other part of Canada. The surface-soil in this district is generally sandy, and covered with pine; but it rests, at a depth of 10 to 18 inches, on a clay often blue in colour, and burning to a white brick. Such a soil is easily tilled at first, and has the means of permanent improvement below.

Wheat is still the surest crop in Canada West, though, in the lower or moister grounds, it is very subject to rust. This has been particularly the case during the last ten years, hundreds of acres having sometimes been left uncut on account of this disease.

Winter wheat is preferred, sown in September, because it leaves less to do in the short spring, is ripe a fortnight earlier than spring wheat, and brings 3d. to 6d. currency a bushel more in the market. It is this winter wheat which the Rochester and Oswego millers covet, because they can get it earlier in the season than the western wheat, and because the flour it makes is less apt to sour at sea.

The best samples of wheat are grown on a belt of some twelve miles broad, which skirts the lake from Niagara round as far as the town of Cobourg, which is about a hundred miles west of Kingston. The land on this belt is rolling, and generally light, loamy, and capable of being ploughed with light horses. Beyond this belt, in every direction, wheat is more subject to rust. Winter wheat cannot be grown with equal certainty, and spring wheat, therefore, is generally sown. From what I have heard, I think it not unlikely that thorough-drainage may eventually cure all this.

The whole produce of Canada West, in 1828, and the average yield per imperial acre, are represented in the following table:—

	Area in cultivation. In acres.	Total produce. In bushels.	Average produce per imperial acre.
Wheat,	593,695	7,558,773	12¾ bushels.
Barley,	29,324	519,727	17½ ...
Oats, .	285,571	7,055,734	24¾ ...
Rye, .	38,452	446,293	11½ ...
Indian corn,	51,997	1,137,555	21¾ ...
Buckwheat,	26,653	432,573	16¼ ...
Potatoes,	56,796	4,751,231	84 ...

Though the largest breadth of land is under wheat, it will be seen by the above numbers that oats and potatoes are staple articles of food. This arises, in part, from the climate being more generally propitious to these crops, and in part from the large proportion of persons of Scottish descent who are found among the inhabitants of the province.

The wheat midge has not yet been sensibly felt in Canada West; still, the wheat crop is by no means so sure as it used to be, and, as one consequence, larger numbers are returning to Indian corn, which, twenty years ago, used to be cultivated in much larger proportion than now.

But it cannot surprise any Old Country farmer that the wheat crops should become less valuable, even on the best of land, where so little has been done to conserve the natural capabilities of the soil. In Prince Edward's District—a peninsula lying immediately to the west of Kingston, between the Bay of Quinté and the lake—the land has, in some places, been cropped with wheat for fifty years, without any other manure than a ton of gypsum a-year applied to a whole farm. In other places, a similar system has been followed. Can it be wondered at, then, that here, as elsewhere—from this among other causes—the wheat-producing region should be gradually retiring inland, or farther to the west; and that, as in the State of New York, the agriculture of the whole province should, by degrees, be materially altering? It is the multiplied and prolonged modes of procedure of its individual citizens that, in agriculture as in the arts, ultimately determines the nature and extent of the relations, economical and commercial, which one country bears to another. These, in the case of the North American provinces and States, are leading to a condition of agricultural production which is of the greatest possible interest to British land-owners and land-cultivators.

Little knowledge of improved agriculture has hitherto been diffused in Upper Canada; and it is, as yet, among practical men, held in little esteem. In revenge, the farming class are not, as a body, regarded with much estimation by the other classes of society. They do not assume their proper position among a community where, if they only knew how to use it, all political power is, in reality, in their hands. The knowledge which they despise would be the means, not only of enabling them to wield this power, but of placing themselves in that position in the social scale from which their present contempt of knowledge debars them. Still, the efforts now making cannot fail to do good. The thousand pounds given in prizes at Kingston had the effect of stimulating many, and of awakening some; while the things exhibited in the show-yard must have led many, on their return home, to look with a new eye upon their own stock and tools and produce. The Society publishes, as yet, no Transactions; but a weekly Journal issued under their patronage promises to spread much useful information. And the lectures delivered from time to time by their Secretary, in the country districts, must lead to thoughts of change, and to improvements in practice; while the introduction, now in progress, of scientific agriculture as a branch of instruction into the Canadian colleges, and of elementary lessons on the principles of agriculture into the schools of the rural districts, will lay a solid foundation for those more general ameliorations by which the practices of the next generation are to repair the evils produced by those of the past.

Among the sources of evil and retardation to Upper Canadian agriculture, the President, in his address, stated the occupation of too much land by individual farm-proprietors to be one of the most injurious. This is especially the case in reference to the descendants of what are now generally called the U. E. Loyalists. At the close of the first American war, many persons emigrated from the United States, not only to Nova Scotia, but also to Upper Canada. They were kindly received, and liberally dealt with by the British Government. All the land on the St Lawrence above the French settlements, up to and around the Bay of Quinté, was divided into townships; the settlers called the *United Empire Loyalists* were placed upon it.

Farming utensils, building materials, and two years' provisions were supplied to them; and, besides the land given to themselves, allotments of 200 acres were granted to each of their children on attaining the age of twenty-one years. This has thrown into the hands of persons of small capital, and little agricultural knowledge or pecuniary means, larger tracts of land than they have been able beneficially to cultivate. Had those grants been made in farms of 100, or even 50 acres each, it would, in the judgment of the President, have been better for the province, and for the actual condition of its practical husbandry.

This, we should say at once, might probably have been the case. But the argument appears to have a better foundation than could be gathered by a stranger as the words fell from the lips of the speaker, when we consider the extent of land granted to the U. E. Loyalists, and how little has, as yet, been done to a large proportion of that which was assigned to their children.

In the authorised Statistical Report of the Canadian Board, (Montreal, 1849,) these lands are described as consisting of:—

	Acres.
Located, . . .	150,800
Unlocated, . . .	321,950
Other lands, . .	2,734,239
Total described, .	3,206,987

I do not know exactly what distinction is intended between the second and third quantities of these lands in the above enumeration; but if it be the case that, of the 3¼ millions of acres granted, only 150 thousand are located—by which I understand actually under cultivation—it does seem as if the original benevolent intentions of the Home Government had been greatly interfered with, and rendered abortive.

A large proportion of the later grants, I believe,

never came into the possession of those for whom they were intended. The claims, prospective rights, or warrants of the children of the loyalists, for 200 acres each, were in great numbers transferred to other parties for small sums of money, and thus came into the hands of persons—often speculators—who have not themselves hitherto possessed the ability or the intention to bring them into cultivation.

Such unlooked-for occurrences as this are to be classed among the other unanticipated consequences which have followed from grants of land made in Canada, often with the best intentions—consequences to be regretted, and which may retard, but, it is consolatory to think, cannot prevent the growth and prosperity of this rapidly rising colony.

The land in Upper Canada is generally cultivated by its owners, as in the United States. In the Gore District, which lies at the head of Lake Ontario, and contains land of the best quality, only about one in twenty is let to a tenant. In the newer settled districts, the system of letting in shares is most common; if the landlord gives only the land, he has a third—if he finds stock also, he gets two-thirds. In the older settled districts, money-rents are common, and leases of seven years are granted, with restrictive conditions as to cropping. Good wheat-land, not within ten or twelve miles of a town, lets at two dollars—about 2½ bushels an acre.

In the table given in a preceding page, it will be seen that the quantity of barley grown in Upper Canada—half a million of bushels—is comparatively small. The produce of this grain is not, however, as in Great Britain, any test of the amount of fermented liquors, and especially of ardent spirits, which are manufactured and consumed. I have already stated, in reference to the Indian corn of the Western States, that distillation was one of the outlets by which the excessive produce of this grain had hitherto been disposed of. Cincinnati, in Ohio, of which I have spoken as the great centre of the *packing* business, is also the great centre of the whisky manufacture. It is the great whisky mart of the West, and probably larger stocks of whisky are to be found at one time in that city than in any other market in the world. The whole quantity produced by the distilleries of this city, or brought into it from more or less distant distilleries, is about 1000 barrels, of forty gallons each, per day, or 14,500,000 gallons in a year. The quantity shipped off in the state of whisky—chiefly down the Mississippi — is estimated at 11,000,000 of gallons, while about 1,000,000 gallons more are converted into alcohol, and disposed of to the Atlantic States. All this whisky is manufactured from Indian corn; and even for the mashing, barley is not necessary, as sprouted Indian corn makes a malt as serviceable to the distiller as that from barley.

In Canada, which is not so much of an Indian-corn country, this grain is also used in the distilleries, although not so exclusively as in the Western States, where this grain is a drug. I had the opportunity of conversing with the intelligent and enterprising owner of a large distillery in the neighbourhood of Kingston, who informed me that he used chiefly rye and Indian corn, but sometimes *pease* also—all ground up together. Two bushels of barley malt were sufficient for a bushel of crushed rye—Indian corn requires four bushels to one. When barley is scarce, a larger proportion of rye can be used. I was most interested in the use of pease, which, from their composition, one would not expect to be well fitted either to give a good sample or a large return of spirit. He informed me, however, that the yield was tolerably good, but the quality inferior to that from Indian corn—the main objection being, that the spirit carries the flavour of the pea along with it.

Though Kingston possesses, in its happy position, a certain assurance of great future prosperity, its progress was somewhat checked by the removal of the seat of Government to Montreal, upon the union of the provinces. Placed at the head of the navigation of the St Lawrence, at the junction of the Rideau Canal with Lake Ontario, and with direct access to the commerce of the States and upper lakes by steam-boats and railways, it will grow with the general growth of Canada, especially with the settlement of the basin of the Ottawa, and the increase of the carrying trade of the great river, till it will compete on at least equal terms with Rochester and Oswego, on the south side of the lake. It is not, as some fancy, feverish energy and over-speculation that are required, but a patient trust in the natural development of the resources of the country, and a prudent and cautious use of the new opportunities of advancing it which every succeeding year presents.

Sept. 22.—Leaving my kind and hospitable friends in Kingston, I embarked for Montreal at 7½ A.M. We had neither rain nor fog in sailing among the Thousand Isles, but the absence of the sun robbed this part of the voyage of half its beauty. I was reminded of the Ten Thousand Islands of the Swedish Lake Maeler, and of the less numerous islets of our own Loch Lomond, as we glided rapidly down the stream; but not a gleam of sunshine descended to give the Canadian scenery the bright sparkle which I have seen lending so much joyfulness to these European lakes. A quiet beauty, nevertheless, suffused the river, and, with agreeable and instructive society on board, the day passed pleasantly. Darkness had already come on for more than an hour before we stopped for the night at Coteau-du-Lac, at the foot of Lake St Francis, and 160 miles below Kingston. We had, during the day, descended several rapids, which can only be passed by ascending vessels through the short canals which have been constructed for the purpose, near the banks of the river. But the most formidable of the river rapids were yet to come; and it was to obtain daylight for the passage of these that we pulled up at the foot of Lake St Francis.

Sept. 23.—At four in the morning we were again under weigh, and most of the passengers on deck, to witness the running of the three formidable rapids, which occur within the next sixteen miles. The descent was very interesting. The rapid current, the often narrow channel, and the care in steering, all told of difficulty in the passages; and when one looked at the large ship, dodging, as it were, among the shallows and headlands, it appeared really wonderful that accidents should so rarely happen. At the foot of the cascades we entered Lake St Louis, where the Ontario from the north falls into the St Lawrence; and at seven we reached Lachine, in the island of Montreal. Here most of the passengers landed, and proceeded by railway nine miles to the city. The really dangerous rapids, however, were still below us, and as the boat was about to pass them, I and a few others remained on board, with the captain's permission, till the boat arrived at Montreal.

This was certainly the most striking part of the voyage; and it is one which a stranger visiting Montreal ought not to allow himself to be prevented from performing. Of the rapids between Lachine and Montreal the most formidable and dangerous, is that of the Sault St Louis. The descent of this rapid, in so large a vessel, created in my mind a feeling of surprise. In descending the Tobique, in my bark-canoe, with a single Indian polling and fending off, in quick and narrow and rocky rapids, I could not help admiring the nice tact, the instinctive perception as it were, with which a gentle touch of the pole on the threatening rock, at the proper moment, kept all safe. Here, on the St Lawrence, the same tact appeared, but with a greatly superior intellectual skill, in handling and guiding a large boat with a heavy cargo through a crooked channel, where the slightest oversight, for a single moment, would cast all upon the rocky shallows.

Let the reader fancy to himself a ledge of rocks running across the river, over which the water has a distinct fall—to the eye appearing to be somewhere between six and ten feet—into deep water below. Through this ledge is a narrow channel of deep water, where the rock has been torn away, and through which the river rushes with great velocity. Below this ledge, at a short distance, is a second ledge of rock, over which the water falls, and through which, as in the case of the first, a natural gap or sluice-way exists. Between these two ledges deep water exists, but the openings of the two are not opposite to each other, or in the same line. You must descend the one, then turn sharp in the deep water along the foot of the first ledge, and at the proper time turn sharp again to go through the other. The channel is a true zigzag, and to sail along this letter Z in the face of a strong current, and a heavy pressure of water, requires a degree of skill and coolness in the captain, and of mobility in the ship, which it requires a little consideration fully to realise. Four men at the wheel, and six at the tiller, to guard against accidents, steered us safely down; and it was beautiful to see with what graceful ease and exactness the prow of the long vessel turned itself to suit the sudden turns of the rocky channel. We reached Montreal about nine o'clock, soon after which, a pelting rain came on — the first serious fall of rain I had yet encountered on the American continent.

The approach to Montreal from the river reminded me of the approach to Leith from the river Forth. The town of Montreal on the river bank, and the hill of

Mont Royal rising behind with a faint resemblance to Arthur's Seat, sent the heart home to more familiar scenes, and almost secured beforehand for the stranger-city an interest in its affections.

In descending the St Lawrence from Kingston, the somewhat naked and rocky limestone country of that part of Upper Canada continues, till we have passed the Thousand Isles. Below this the banks are less rocky, and most of the way down to Montreal consist of a light-coloured drift, which yields in general, I should think, only an indifferent soil. This drift rests upon, and is probably in great part formed from, the Potsdam sandstone and calciferous sand-rock of the New York geologists—being the lowest portions of the Lower Silurian rocks. These rocks, where they occur in other parts of North America, produce in general inferior soils.

The most interesting geological fact, bearing upon the practice of agriculture, which fell under my observation in this part of my tour, is the occurrence over a large part of Canada of a deposit of metamorphic limestone, which is unusually rich in phosphate of lime. This limestone is subordinate to, and interstratified with, beds of porphyritic and syenitic gneiss, which form a long ridge of high-land, dipping beneath not only all the Silurian strata, but also under the copper-bearing beds of Lake Superior, which are beneath the Silurian. Both the limestone and the gneiss are probably highly altered members of the older Cambrian series.

This ridge of altered rocks extends, as a prolonged high-land, in a north-east and south-west direction, from the west of Labrador to the Ottowa, running nearly parallel to the St Lawrence, and at a distance north of that river of from twelve to twenty miles. Near Bytown on the Ottowa the limestone appears in great force, and from that point the ridge of mixed rocks ranges nearly due west to the shores of Lake Huron.

Near the point where it crosses the Ottowa, a branch of this formation forks off towards the south, spreads over a considerable extent of country between the Ottowa and the head of the St Lawrence, crosses this river at the Thousand Isles, among which the syenitic rocks prevail, with intermixed crystalline limestones, and passes into the northern counties of New York, where it is extensively developed. It is there coloured among the primary rocks of the State, in the published geological maps of Mr Hall and Professor Emmons.

This rock, like the altered limestones in most other localities, contains imbedded in it various simple minerals in greater or less quantity; and among these apatite, or phosphate of lime in grains and green crystals, is sometimes very abundant. Mr Logan, Provincial Geologist for the Canadas, in his Report for 1845-46, p. 94, has mentioned several localities where the mineral phosphate of lime is especially plentiful;* and Mr Hunt, chemist to the survey, with whom I had the pleasure of conversing upon the subject, assured me that in many places this mineral formed a tenth part of the whole rock.

One economical fact is certain—that the existence of such a limestone is of undoubted value to the neighbourhood in which it exists, where it can be readily quarried and burned for lime, to be used in agricultural operations; and that it is of equal value as an article of export for agricultural purposes, where facilities for shipment or other cheap means of transport exist. Such a limestone rock, in most easily accessible parts of Great Britain, would be as sure a source of permanent wealth as a mine of Californian gold.

Another economical point is worthy of inquiry. Does this mineral phosphate, in any of these localities, occur

* At Blasdells Mills, on the Gateneau, at the Calumet Slide, and above the head of Moor's Slide, near the line between Ross and Westmeath. Probably many other localities are now known.

in masses so large, or so readily separable from the common limestone, that it could be economically extracted, and brought in a pure state into the market? If so, it would prove valuable as an article of shipment to Europe, and would provide another available resource to the high-farmed lands of Great Britain. Upon the exhausted wheat-soils of Canada, properly prepared and applied, its use would be invaluable. An inquiry into this point is deserving of the attention of the Canadian Legislature, with a view to the good of the province; and of individual landowners along the outcrop of this rock, with a view to their own individual profit.

I think it the more likely that some localities may be found in Canada where this mineral phosphate will present itself in sufficient quantity to admit of being profitably extracted, because, during my subsequent residence in the State of New York, I was assured by Professor Emmons of Albany, one of the State geologists, that he had met with it in several places in that State where he thought it might be so extracted. In the white metamorphic limestones of Essex, Jefferson, and St Lawrence Counties, into which, as I have already said, the Canadian limestone extends, he had so observed it; and at Rossie, in the last of these counties, he thought a man might in some places pick out a hundredweight a-day. Among the magnetic iron ores also, in the township of Peru, near Lake Champlain, he informed me that it sometimes occurs in equal bulk with the ore itself, and that, by washing or other mechanical means, a ton a-day might be collected in some localities.

In the interest of general scientific agriculture, independent of individual or local profit, it is desirable that the accuracy of such statements, and the possible availability of these and similar deposits, should be speedily investigated.

Speculations have been hazarded at various times in regard to the origin or source of the crystals of apatite (phosphate of lime,) of graphite (plumbago,) garnets, and various other minerals, which are met with in so many countries intermingled with the metamorphic or crystalline limestones. But the origin of all these is now easily intelligible. It is certain that this crystalline character is the result of the action of heat long continued. But the assumption of this crystalline character implies a power of movement of the particles among each other, which, in fact, is seen in many cases in unmelted bodies—as in the annealing of glass and metals, and in the tempering and converting of iron or bronze—to take place where they are kept for a prolonged period at an elevated temperature. It is certain, also, that particles of a like kind have a special attraction for each other—a tendency to draw towards one another and cohere, when circumstances are such as to admit of their moving among themselves, or among the particles of other matter with which they may be mixed. And, thirdly, it is certain that, when several substances which incline to unite with each other are present in a mixture in which circumstances admit of a movement among the particles, they often unite to form definite chemical compounds, exhibiting, more or less, well-defined crystalline forms.

Now it is known that stratified limestones, when deposited, are rarely free from admixtures of earthy matter, which contain the constituents of garnet, chondrodite, hornblende, &c. When these limestones are subsequently exposed to the long-continued action of heat, the particles of the rocky mass arrange themselves in crystalline forms, while the earthy matters unite to form the simple minerals (garnet, &c.,) which can be most readily produced out of the substances of which they consist, in the proportions in which they are actually present. Hence the minerals produced differ in quantity, in kind, and in relative proportions, according to the quantity and nature of the impurities which the limestone contains.

Again, all limestones, almost without exception, contain the remains of animals or vegetables, or both. In the former they are often very rich, and in these phosphate of lime always exists in sensible quantity. When the limestone is changed by heat, the animal and vegetable substances are at the same time entirely decomposed. All that is volatile escapes; while the earthy phosphate, being fixed, remains intermixed with the limestone. But while the particles of carbonate of lime attract each other, and form crystalline marble, those of phosphate of lime also attract each other, and, although sparingly mixed up with the limestone, gradually approach each other, and finally cohere into crystalline grains, and regularly crystallised forms. In this way the phosphate, which was intermixed with the rocky mass, perhaps in almost inappreciable proportion, becomes collected together into sensible masses in particular places. And if the rock be one which, like some of our still unchanged limestones, is unusually rich in animal remains, or in mineral phosphates as a mass, the quantity of the separated crystals will be great in proportion. Hence their comparative abundance in some of these white limestones, and hence also the reason why the occurrence of them may be looked for, in some localities, in sufficient quantity to admit of their being economically extracted for use in agriculture or the arts.

Lastly, while the volatile parts of the animal and vegetable matters contained in the limestone—those which often give to limestones a bituminous character—escape under the influence of heat, a portion of the more fixed charcoal (carbon) remains, in the crystallised form of graphite or plumbago. Hence, the deposits of this substance, more or less extensive, which are frequently met with in metamorphic limestones—varying in amount with the quantity and kind of organic matter which the rock originally contained, and with the intensity and continuance of the heat to which it has been exposed. In some places in Canada, and in the State of New York, therefore, it may still be found in sufficient quantity to prove an invaluable source of mineral wealth.

If what is above said, in regard to the phosphate of lime, be received as a satisfactory explanation of its origin, it will follow that crystalline limestones in which this mineral is found must not only be metamorphic, but must have been deposited in the stratified form, and have once contained the remains of fossil animals in very considerable quantity.

I may here, in connection with this metamorphic limestone of Canada, mention three facts observed by Mr Logan,* which are not only interesting in themselves, but which bear upon the important point in chemico-geological theory—the supposed origin of such limestones, to which I have just alluded. These facts are—*First*, That while the gneiss above or below the limestone exhibits regular stratification and even lamination, the limestone itself will at times display contortions of the most complicated character, and which increase in importance with the thickness of the bed of limestone. *Second*, That when this thickness is great, beds of gneiss, of even a foot in thickness, will be bent, folded, and broken, and fragments—sometimes very large—of the gneiss will be surrounded by the white limestone. *Third*, That, in one instance, the bed of limestone had an uninterrupted connection with a mass of the same, which filled up a crack or fault in the gneiss, at right angles to the general direction of the strata. In the large, the limestone and

* *Survey of Canada Report* for 1845 and 1846, p. 43.

the gneiss were interstratified regularly and distinctly. How, then, came these anomalous appearances? They are regarded by some as proofs of the theory that the crystalline limestones, like the granites which have been fused, are of igneous origin; that they have been melted, and in this state have been elevated from beneath, and injected into the anomalous situations in which they are occasionally found. Without denying the possibility of such fusion and injection, there is no occasion, I think, to have recourse to this supposition with the view of explaining the phenomena hitherto observed. The most strenuous supporter of this igneous origin of the crystalline limestones in North America is Dr Emmons of Albany—a man of much learning, who has enjoyed many opportunities of personally observing such rocks *in situ*. My attention was drawn to the subject by this gentleman during my stay in that city; I shall, therefore, if my space permit, return for a little to the consideration of his views in a subsequent chapter.

CHAPTER XI.

Montreal.—New churches.—Ruins of the Parliament House.—Scotch farmers in the Island of Montreal.—Soil near Lachine.—Its produce per acre.—Cultivation of hops.—Price of land.—French Canadian farms and farming.—Bad farming at home.—Bad farmers forced to emigrate.—Clerical obstacles to the settlement of Protestant farmers in Lower Canada.—Apples and cider of the island.—Hedges of English and American Thorn.—Valley of Lachine.—Importance of a better practical husbandry in Lower Canada.—Interest of an established clergy in promoting its introduction.—Interest they have in the land.—Yet progress has generally been slowest where the interest of the clergy is the greatest.—Importance of agricultural instruction in the country schools, especially in *agricultural principles*.—How they should be taught.—Desire to introduce such instruction in Lower Canada.—Patronage of the Roman Catholic clergy.—Excursion to St Hilaire.—St Lawrence and Atlantic railroad.—Winter shelter for cattle.—Mode of making butter.—Maple sugar manufacture in Canada and the adjoining States.—Mode of procedure.—Produce of single trees.—Profit of Maple-groves.—Soil of the valley of St Lawrence.—Tile-drainage upon its stiff clays.—Pigeon or stone weed, its prevalence.—What its history teaches.—Inferior breeds of pigs and cattle.—Bellœil Mountain.—Pilgrimage Stations.—Beautiful view of the St Lawrence flats.—Exhaustion of this formerly fertile region.—Seignorial tenure of land.—Reserved rights of the seigneur.—Reserved rents a grievance.—Sherbrooke.—Lands of the "Canadian Land Company," in the eastern counties.—Their progress and present inconveniences.—Tile-draining at Montreal.—Voyage to Quebec.—The Ottawa River and District.—Its rising importance.—Proportions of British and French in Montreal.—National, political, religious, and municipal parties in the city.—Difficulty in satisfying such a population.—Wisdom in changing the seat of government.—Why the British members from Upper Canada voted for the Rebellion Losses Bill.—Explanation of one of their number.

Sunday, 23*d September*.—I attended morning service in an Episcopal chapel in Montreal. It was well performed; but the congregation was very small. In its ecclesiastical concerns, the city appears to be in a very prosperous and thriving condition. A new Roman Catholic cathedral, and a palace for the bishop, have been of late years erected. The cathedral is chiefly remarkable for its size and internal capacity. It is so fitted up with pews as to afford sitting-room for ten thousand people! New churches belonging to other denominations also, and all handsome, are springing up in various well-selected situations.

As I went along the streets, evidences of the prevailing political excitement presented themselves everywhere. Among others was a large placard on the walls, headed "Rebellion Rewarded," and containing extracts from the speeches of Lords Stanley and Lyndhurst in the House of Lords, on the subject of the recent Compensation Bill. I could not help regretting that opinions should have been expressed by influential men in the Home legislature, which those who were now declaring themselves enemies of British connection should be able to quote with something like a show of reason, in justification of their illegal violence.

I visited the ruins of the Parliament House. It had been a fine massive building, constructed of the blue Trenton limestone, which is much employed as a building stone in Montreal. The outer walls were still standing; but though massively built, it appeared to be entirely incapable of repair. It is a defect of this stone, otherwise excellent for building purposes, that the united action of fire and water upon it causes it to crack, and fly into numberless splinters. Hence the walls of the Parliament House, and those of a large hotel lately burned down in the city, are shattered and splintered in every direction. This is an evil to which walls of brick, sandstone, granite, or hardened slate, are not so subject.

Monday, Sept. 24.—The wind was blowing cold this

morning, and I began to feel my summer clothing too light for the Canadian autumn. After breakfast, I drove out to Lachine with a Mr Somerville, a practical Scottish farmer, long settled in the country, who has a farm of 300 or 400 acres on the banks of the St Lawrence, opposite the rapids of Lachine. He and his neighbour, Mr Penner, are farmers of the old Scotch school, who bid you " lay the land dry, then clean and manure—make straight furrows, clean out your ditches, take off the stones, and plough *deepish*." With these good mechanical principles, industriously carried out, they have greatly surpassed the French Canadian farmers; and with the possession of good Ayrshire stock, and the growth of a few turnips, and of mangold-wurzel, which does well even with the early winters of Lower Canada, they have raised good crops, extended the arable land of their farms, and kept up its condition.

The soil on this part of the island of Montreal, which lies low, and along the river, is of a blackish colour, generally very rich—of a loamy character, and easy to work. It is drained by open ditches and cross-furrows. Tile-draining, hitherto untried, is indicated by the local circumstances of soil, climate, and physical position; and although here, as in the State of New York, the cost may appear large when compared with the total value of the land, and the increase of price which, after tile-draining, would be obtained for it in the market, yet, if from the cost be deducted the annual outlay which must be incurred to keep the ditches and cross-furrows open, the actual expense of the permanent tile-drainage will rapidly disappear. When a man settles on such land, therefore, as requires the maintaining of open ditches—with the view of retaining it, say only ten to twelve years—he will, in most cases, find his pecuniary profit greater at the end of the term, although the price he then sells his land for should really be no greater, in consequence of the drainage he has performed. Here, as in New Brunswick and the Eastern States of the Union, I find it was a disputed question whether money is to be made by farming, where all the work is done by hired labour; that is, whether the Scotch and English system of large farming, or the class of large farmers, can be successfully introduced into the province.

It is conceded that a man with 100 acres in cultivation, doing one-half the work by the hands of his own family, and employing hired labour to do the rest, may make both ends meet; but if a larger farm is to be worked by the same home-force, with a larger number of hired labourers, it is a question whether it can be done, in average years, so as to pay. This doubt arises not merely from the high price, but from the alleged, and I believe real, inferior quality of the agricultural labour—chiefly Irish—which a farmer is able to secure.

The island of Montreal has been long celebrated for its fertility, and, from its production of fine fruits, has been called "the garden of Canada." The front-land, along the river, over which I passed, is very good, producing, per imperial acre, from 20 to 35 bushels of wheat, from 40 to 60 of oats, and of Indian corn, though not much cultivated here, from 40 to 50 bushels. The value of this land is, on an average, about £20 currency, or £16 sterling, per acre, when it is in a good state of cultivation, and has good buildings upon it.

Mr Penner's farm is, for the most part, very superior land. From 40 to 50 acres of it are in hops, which thrive well—produce, on an average, from 800 to 1000 lb. per acre, and are a profitable crop. Here, as in our own hop-grounds, and in those of Flanders, they require high manuring; and thus, as a general article of culture, they are beyond the skill of the manure-

neglecting French Canadians, and the equally careless British and Irish emigrant settlers. This rich hop-land is worth £40 an acre.

At a distance of twenty or thirty miles from the market of Montreal, *good land* can be bought for £4 to £7 an acre; but the buildings are generally bad. If these happen to be good, the price is higher. The farms in this district usually run in long stripes, the breadth of 3 acres in front, and from 30 to 50 back, or from 90 to 150 acres in all. The average produce on the farms of the Canadian French is not more, as I was informed here, than 8 to 10 bushels of wheat per acre—and for this grain the soil is becoming poorer—and 20 bushels of oats. They grow little hay, except on the natural meadows, from which they reap 1 to 1½ tons an acre. The system is to divide the farm in two, lengthwise, and to crop the one and pasture the other alternately, without sowing down—merely grazing the weeds that spring up.

This neglect of grass-seeds may be considered as a fair indication of a low state of practical husbandry, in nearly every country which is blessed with a moderately moist and temperate climate. It is far too general in North America—the Provinces and States alike. Even among our home-farmers, it is to be observed much more frequently than those persons who never leave the high-roads in their agricultural travels would readily believe.

Indeed, if we go a little out of the beaten track, we may find, either in England or Scotland, all the vices of American farming. Our home-farmers, indeed, may be said to be the parents of them all. We have among us still numberless farmers of the old school, possessed of deep-rooted prejudices, who refuse to advance and change their methods. It is chiefly such men who, during the last twenty or thirty years, while those who advanced with the times were generally prosperous, have, one by one, been driven from their farms, and forced to emigrate. Having sacrificed themselves at home to their prejudices, they bear them religiously beyond the Atlantic, and transmit them as heir-looms to their descendants.* The changes in our corn-laws will not, it is to be hoped, now send out any of our better men.

With such men, holding at least a considerable share of the land, it is not surprising that bad farming should be found in North America, even among the Anglo-Saxon race; and if improvements are introduced slowly among us, they cannot be expected to advance with a less languid step among them.

It is believed that the introduction of British settlers into Lower Canada would improve the rural industry of the French population; and, in so far as the example of a more patient and energetic blood goes, this might possibly be the case. But, in addition to the unwillingness which the British Protestant emigrant feels to place himself in the midst of a people who speak a different tongue, and belong to a different religious denomination, there is another obstacle to this admixture of races, arising out of the law of tithes, of which I was not aware until my friends explained it to me here to-day.

Before the British conquest, the Roman Catholic clergy were, by law, entitled to the tithes of all land—one twenty-sixth part of the produce being the legal due of the priest of the parish. But by what is called, I believe, the Quebec Act, they are now permitted to demand tithe of persons of their own persuasion only, Protestants being exempt. Hence, every transfer of land from a Roman Catholic to a Protestant proprietor, is a money-loss to the Romish Church, and a money-inducement is held out to the priest of the place

* See, for instance, the state of farming in Lancashire even now.—*Royal Agricultural Journal*, vol. x., part 1.

to throw obstacles in the way of such transfers. This is a temptation by which the best of men among the Roman Catholic clergy may be insensibly biassed, and to which the law ought not to expose them.

Fruit, and especially apples, grow well on the low-lying black land I visited to-day; and orchards are numerous around Montreal. At Mr Penner's, I saw some fine apple-trees, and gathered some excellent fruit. Among other eatable varieties, the *pomme grise* is one which is highly esteemed as a good bearer and of high flavour. Some of the valuable rennets also, such as the large white Canadian, called also the large English rennet, are said to be of Canadian origin. Besides table fruit, cider-apples also are grown; and in addition to those of his own orchard, Mr Penner buys and crushes those of the neighbouring orchards.

For the first time in Canada, I saw here some hedge-rows of our English thorn, *Cratægus oxyacantha.* The mice are said to be their greatest enemies. When the snow is on the ground, they often destroy this and other valuable trees, by gnawing off the bark for food—as hares and rabbits, in severe winters, do with us. I saw also some very nice and well-kept fences of the American thorn—the *Cratægus crus-galli*, or cockspur thorn, I believe.* No other objection was stated to the general introduction of these hedges, except the trouble and expense of keeping them in repair.

Before leaving Mr Penner's farm, I ought to mention, in connection with the high-manuring of his hop-lands, that he has a bone-mill upon his farm—that there is another at Montreal; but that, as I was informed, scarcely any true Canadian has as yet seriously thought

* In the State of New York, four native species of thorn are known, of which this is considered the best suited for hedges, and is most frequently employed for this purpose. I am not aware if all these species are natives of Canada also.

of employing such a thing as bones, even for the manuring of his worn-out wheat-lands.

I returned from Lachine, by the way of the valley—apparently an old channel of the St Lawrence, or of the Ottawa—along which the ship-canal and the railway have been conducted. There is rich flat meadow-land in the bottom of the valley, only partially drained; and some tracts of rich alluvial soil, dry enough to be submitted to arable culture, and productive of excellent crops. The upland, also, is generally of good quality, and well-cultivated farms are not unfrequent. On one of these I met with a warm and kind reception from Mr Evans, the Secretary of the Lower Canada Agricultural Society, editor of their Journal and Transactions, and the author of a valuable treatise on agriculture, adapted to the climate and productions of Canada. Both on this and on subsequent occasions, I was indebted to the kindness and attention of Mr Evans, and was obliged to him for much valuable information.

Sept. 25.—Among other persons whom it gave me pleasure to visit this morning was M. Morin, President of the Legislative Assembly, and of the Lower Canada Agricultural Society; and M. Villeneuve, Principal of the Roman Catholic college. Both of these gentlemen expressed a strong desire to promote the introduction of a better system of practical agriculture among the inhabitants of Lower Canada; and no other kind of instruction—I may say, from all I afterwards saw, no other gift which can be bestowed upon this people—seems likely to be productive of more material good to the province.

Looking at the relation which an established clergy bears to the agriculture of a country, in one of the aspects in which it naturally presents itself, it would appear as if no class of men ought to be more anxious to promote agricultural improvement, to remove obstacles out of its way, and to diffuse that kind of knowledge by which it is to be most rapidly advanced. Entitled in most countries to a certain fixed share of the produce of the land, the larger that produce can be made, the greater the revenue the clergy must yearly receive. And yet experience seems to show that it is precisely when, as in Scotland, the established clergy have the least interest in the amount of produce yielded by the land, that agricultural improvement has most progressed; while it has remained most backward, also, in those Roman Catholic countries in which their interest has remained the greatest. Every one, in fact, knows how the tithe question has impeded rural improvement in countless localities, even in England; and how the tithe-commutation measure has been introduced, in the hope of removing the obstacles it presented, not more to rural peace than to rural progress.

It is not difficult to understand how such obstacles should actually arise out of what, at first sight, appears likely to promote agricultural improvement; how a diversity of interest should exist between the cultivator and the tithe-collector; and how human nature should stubbornly, though foolishly, refuse to adopt new methods which would be more profitable to the farmer himself, simply because they would at the same time be a source of profit to another, who incurs none of the additional labour, anxiety, or expense.

There is, however, an indirect method by which improvements are certain to be brought about—slowly perhaps at first, but largely and generally in the end. This method is the general diffusion of knowledge bearing upon the practice of agriculture. It is not by prescribing new methods to old men—by staking our chances of success on the hope of overcoming the prejudices of the most prejudiced class of society. It is by instilling into young and unprejudiced minds the principles according to which all rural practice ought to be regulated, that future practice will be most certainly made better. This can be done at little or no expense; and in regard to imparting this knowledge, there can be no opposition of interests between the clergy and the rural laity. In most places, the parents will regard the new instruction as a boon to their children, and will be proud of their knowledge. They will, in most cases, also be delighted to see their children apply this knowledge under their own eye; and if they themselves refuse their assent to the introduction of this better culture, it is sure to be seen on the farms to which their sons succeed. To all, therefore, who have an indirect interest in the better tillage of the land by those who hold it, the diffusion of such knowledge among the young in our rural districts at home, as well as abroad, ought to be a chief concern. Especially in our home islands, now when rents are falling, if knowledge can by possibility be made to keep them up without diminishing the comforts or reasonable profits of the farmer, it ought to be liberally, and with a ready hand, scattered among the children of the people.

To be generally available, however, the mode in which this is done should be easy, short, inexpensive, involving little change in the ordinary school-routine, little new machinery, and little interference with the customary school-teaching, in kind or quantity. All this, I think, may be effected, if the eye is kept bent upon the one object—that of instructing the children in *agricultural principles*, and their modes of application. These are comparatively few in number—can be simply expressed, so as to be intelligible to the very young; and can be taught in so short a time as to interfere in no necessary degree with the usual branches of education.

It is difficult to impress this clearly and distinctly either upon the general mind, upon that of teachers themselves, or even upon that of school-inspectors. The principles I speak of are deduced from scientific inquiry—

chemical, geological, botanical, and physiological research. In the expression of these principles, new words—the names, for example, of certain substances familiar to the chemist or botanist—are necessarily employed. These words or names must be understood, if the sentence in which they are contained is to be comprehended—as the child is shown pictures of the horse and the lamb, or is taken to the fields to see these animals, if it is to understand the early reading-lessons in which they are mentioned. But a thing is known by its sensible properties; and as a child at once distinguishes the apple, the potato, the turnip, and the onion, by their form, colour, taste, and smell, so, among the things chemistry deals with—phosphorus and sulphur, oxygen and nitrogen, starch and gluten, must be made familiar to his senses, if he is to understand the meaning of their names. Thus far experimental chemistry is necessary in the teaching of agricultural principles. It must make the words intelligible, but no more is necessary. With the apparatus or book in his hand, however, the master (if the purpose of his teaching be not clear to his own mind) is apt to introduce other unnecessary experiments or scientific explanations, burdening the memory of the boy, and distracting his attention. The school-inspector, also, not distinguishing more clearly the true line of agricultural teaching, sometimes encourages this, by requiring and expecting, at his examinations, a knowledge of purely chemical principles, which are necessary neither to the comprehension nor to the future application of the agricultural principles which are intended to be inculcated. A teacher must learn to resist the temptation to show a pretty experiment, even with the laudable desire of making his pupils see and feel its natural beauty, for the boy will naturally ask, "How does this apply to agriculture?" It requires a clearer head and more considerable knowledge than many teachers of elementary schools possess, to be able to draw the distinct line I speak of, between what ought to be taught and what withheld, and considerable self-restraint to keep within that line, when it is seen. The school-inspector, therefore, at his periodical visit, should sedulously assist, by the questions he puts, in keeping him to the special instruction which is to be given, otherwise this branch of scientific agriculture will be made to occupy too prominent a place in the course of instruction, and will take up more time than is properly due to it, while the pupils will, at the same time, be less satisfactorily or usefully taught.

These remarks I introduce here, as likely to be by no means without their use even among ourselves; for though I have often, in various ways, pressed such views upon teachers and parents in our rural districts, and though as many as 26,000 copies of my little Catechism —which contains all the chemistry, and all the agricultural principles which, in my opinion, are necessary to make the schoolboys of our day the agricultural improvers of the future—have been distributed at home, yet in many cases I have reason to fear that the prospect of good has been marred by a departure from the simple necessities of the case, and the introduction of what is really extraneous matter. For whatever may be said in favour of pure chemistry as a separate branch of school-teaching, it is not included in, and ought not to be unnecessarily mixed up, in *elementary schools*, with instruction in the principles of agriculture.

Upon this subject of instruction in the principles of agriculture in the common schools of the country, I had frequent opportunities of conversing with influential persons in the British Provinces and in the United States, and rarely, I believe, without awakening a desire, or strengthening that which pre-existed, to promote this object, as a means of most certainly bettering the future condition of practical agriculture everywhere.

M. Villeneuve, the Principal of the Roman Catholic college at Montreal, I found already alive to the subject. He had taken one of the college farms into his own hands, with a view to model improvements; and on his library table I was pleased to find a copy of the fifth American edition of my published Lectures. In no country in the world, as my subsequent experience taught me, is the application of a greater amount of knowledge to the soil more necessary than in Lower Canada. Through the primary schools it can be brought to bear upon the practice of the next generation, without interfering directly with the prejudices of the present; and thus, without clashing of interests, the common good of all may be surely and peacefully promoted.

My stay in Montreal was too short to allow me the opportunity of meeting and conversing with the numerous other persons who are interested in the improvement of Lower Canadian agriculture. I was happy, however, to find, that the heads of the Roman Catholic clergy, here and at Quebec, were among the chief supporters of the Agricultural Society of the Lower Province, and were exerting themselves to interest the inferior clergy, so influential among the habitants, in the introduction of a modicum of agricultural instruction into the schools under their charge.

In the afternoon of this day I crossed the St Lawrence with Mr Wettenhall, an influential member of Assembly from Upper Canada, and Major Campbell, of St Hilaire, Secretary to the Governor-general, who had kindly invited us to visit him at his place of St Hilaire, about fifteen miles from Montreal. We crossed the river by a steamboat, and descended a little way to Longeuil, whence we proceeded by the St Lawrence and Atlantic railroad, which has already been opened as far as St Hyacinth, a distance of about twenty-seven miles. This railroad, which will afford the easiest and shortest line yet projected from Montreal to the Atlantic, will reach the coast at Portland in Maine. When it is completed, which is expected to be the case in two or three years, it will not only greatly promote the prosperity of Montreal and of Quebec, to which city a branch is intended to fork off, but it will much assist the city of Portland to compete with the growing neighbour city of Boston, which it is anxious to rival and surpass. To be at the mouth of a long river, or at the terminus of a long inland railway, and upon the Atlantic border, seem, from past experience, to be sure preludes to commercial prosperity in a North American city. It is not surprising, therefore, that New York, Boston, and Portland, should all be anxious to perfect, complete, or shorten their lines of communication with the Canadas and the St Lawrence.

For ten miles we went over a flat country, chiefly of tertiary or post-tertiary light-coloured clays. Here a break-down upon the line arrested our advance, and we considered the most promising way of getting on was to walk the remaining five miles. This gave us an opportunity of seeing to greater advantage the wooden bridge and viaduct, 1200 feet long, which conducts the railway across the river Richelieu. This bridge, though nothing almost when compared with the engineering triumphs of our British railways, is deserving of a visit, both as a difficult work of art, and as an evidence of the enterprise and energy of a young and growing country. We arrived at St Hilaire three hours behind our time, and fully prepared for an unexpectedly late dinner.

Sept. 26.—This morning was wet and unpromising for a rural excursion in a flat and somewhat clayey country, where the roads are bad and walking difficult.

While the sun came out, however, put an end to the rain and dried the land, we were able to inspect the well-finished stables and farm-offices which Major Campbell has erected, and to inspect his dairy and his stores of maple sugar.

I have already, in speaking of the winters of New Brunswick, made some remarks upon the importance of greater attention to the warmth of the cattle, if their condition is to be easily kept up, fodder saved, and profit to be made by keeping them. In this and the lower parts of Canada, similar observations apply to the mode in which the habitants tend their cattle in winter. In Upper Canada, west of Kingston, open hammels are in use for winter shelter, but in Lower Canada this practice is inconsistent with economy. About Montreal, in the winter of 1848, the thermometer, for three weeks together, never rose above zero. To expose cattle to such extreme weather is to sacrifice food. It struck me, therefore, that, in erecting well-constructed warm winter buildings for his stock, the Seigneur of St Hilaire was setting a most commendable example to his tenantry and more wealthy neighbours.

I do not know how far the method of making butter adopted in Mr Campbell's dairy is common in Lower Canada, but here it is made after the manner of what is sometimes called *Bohemian butter*, or of some of the varieties of the *Epping butter* of England. The cream is collected, Devonshire fashion, in the form of clouted cream, by placing the milk-vessel, in which the milk has already stood twelve hours, upon a hot plate till it is nearly boiling, then setting aside twelve hours to cool, and subsequently removing the cream in the usual manner. In this way it is said that a fourth more butter is obtained, and the churning is performed by merely stirring the cream about with a stirrer, or with the naked hand. Of course the skimmed milk is more worthless, and contains only the curd and the sugar of the new milk.*

But the maple-sugar manufacture of this neighbourhood was more interesting, as it possessed more novelty to me. The importance of this industry to the Canadas may be judged of from the fact that, in 1848, there was made, in Canada West, as much as 4,140,667 lb. of maple sugar, or nearly 6 lb. for each inhabitant. In Lower Canada, in 1844, the quantity produced was 2,250,000. If we suppose it now to be 3,000,000, the whole quantity of maple sugar produced in average years, in all Canada, is about 7,000,000 lb. There are imported besides, of West India sugar, about 20,000,000 lb.—so that the home produce amounts to one-fourth of the home consumption.

Maple sugar is also an important article of rural industry in some of the United States. In Michigan, the produce in 1848 was estimated at 1,774,368 lb.; and in Vermont and New Hampshire, it amounted to several millions of pounds. †

Major Campbell is himself a maple-sugar grower to a considerable extent. On his domain he possesses about 12,000 trees, which yield on an average about a pound

* Another variety, which I believe is the genuine esteemed Epping butter, is made by churning alone the cream which rises naturally during the first twenty-four hours. This gives a much more delicately flavoured butter than when all the cream which the milk will yield is mixed and churned together.

† The Report of the Patent Office for 1847 estimated the maple-sugar crop for that year in New Hampshire at 2,250,000, in Vermont at 10,000,000, and in New York State at 12,000,000 lb. But it estimated that of Michigan, in the same year, at 3,250,000 lb.; whereas the returns published by the legislature of that State make it only 1,750,000 for 1848, which was a remarkably good sugar year. I doubt, therefore, that exaggerations may exist also in the returns for Vermont and New York, as given in the Patent Office Report for 1847, p. 85, and therefore I have not introduced the numbers into the text. The estimate for Upper Canada is taken from the Reports of the Canadian Board of Registration and Statistics, published at Montreal in 1849.

each tree. Some trees yield three or four pounds—a pound being the estimated yield of each *coulisse* or tap-hole—and some trees being large and strong enough to bear tapping in several places. Some years also are much more favourable to this crop than others, so that the estimate of a pound a tree is taken as a basis which, on the whole, may be relied on as fair for landlord and tenant. These trees are rented out to the sugar-makers at a rent of one-fifth of the produce, or one pound for every five trees. March and April are the months in which the trees are tapped, and the best weather is when hard frost during the night is followed by a hot sun during the day. In Upper Canada, from its proximity to the lakes probably, the sugar weather is more variable, and the crop less certain than in Lower Canada.

The first sap that flows in April is clear, colourless, and without taste. After standing a day or two, this sap becomes sweet; and a few days after the tree has begun to run, the sap flows sweet. The last sap flows thick, and makes an inferior sugar, called here *sucre de seve.* When boiled carefully in earthen-ware or glazed pots, the clear sap gives at once a beautifully white sugar, and especially if it be drained in moulds and clayed, as is done with common loaf-sugar. When pure white, however, it cannot be distinguished from refined cane-sugar. It is generally preferred of a brown, and by many of a dark-brown colour, because of the rich maple flavour it possesses—a flavour which, though novel to a stranger, soon becomes very much relished. It is an article of regular diet among the Lower Canadians. On fast-days, bread and maple sugar are eaten in preference to fish. In spring it sells as low as 3d. a pound, but in winter it rises sometimes as high as 6d.

In some of the townships of the Eastern Counties—as the district is called which lies between this place and the borders of Maine and New Hampshire—maple-groves are now planted for the raising of sugar. Such groves, in the opinion of many, will yield more profit than any other use to which the land can be put, as beneath the trees an excellent pasture springs up.

The sugar maple, *Acer saccharinum*, forms extensive natural forests on fertile soils, and especially on those of the Niagara and other limestone formations of this region, though I am not aware that it particularly affects soils of a calcareous character. Into these forests, in spring, the sugar-makers plunge, carrying with them a huge pot, a few buckets and other utensils, their axes, and a supply of food. They erect a shanty in the neighbourhood of the most numerous maple-trees, make incisions into as many as they can visit twice a-day to collect the sap, boil it down to the crystallising point, and pour it into oblong brick-shaped moulds. In this way, in the valley of the Chaudière, from 3000 to 5000 lb. of sugar will sometimes be made by a single party of two or three men.

The day having cleared up, we were enabled to take a short drive over the domain, as far as the mountain of Belœil — a lofty isolated ridge, which springs up in the midst of the plain, at a distance of two or three miles from St Hilaire.

The soil which prevails in the valley of the Richelieu River, and over a large portion of this district, is a stiff, light-coloured clay, which used to be rich in the production of wheat, and was esteemed the garden of Canada. It gradually became impoverished however, till in 1835 the wheat-midge appeared, and almost banished the growth of wheat. In 1848 and 1849 the ravages of the insect a little abated, and small fields of wheat arrived at comparative perfection. Here, as in other provinces and States, it has been observed that late-sown wheat is less injured by the midge than early sown—that which is put in during the third week of

May escaping, while that which is sown during the first week is destroyed.

The whole district, from its flatness and the tenacity of its soil, is a fit subject for thorough-drainage with tiles—for the manufacture of which the soil itself affords abundant material. Major Campbell has imported a tile-machine—the first, I believe, which has been seen in the province—and placed it in the hands of a brickmaker near Montreal, who has already made and disposed of many thousands. On a portion of his own land, which he had dried by means of these tiles, Major Campbell showed me beautiful mangold-wurzel; while on the undrained land beside it, scarcely a plant had thriven. Richer clover also had come up on another drained spot, and less of the pigeon-weed, as it is here called, with which this clay land is infested.

There are few of the evils which afflict the practical farmer, in the management of his land and stock, which do not afford illustrations of the money-value of knowledge to himself—and of the economical importance to his country that he should possess it. I may advert for a moment to the lesson taught us by the weed of which I have just spoken.

The *Lithospermum arvense*, Corn-gromwell or stone-weed—called in North America by the various names of pigeon-weed, red-root, steen-crout, stony-seed, and wheat-thief—is said to be a European importation, brought in probably with unclean seed-wheat from France, or Germany, or England. Thirty years ago, it was almost unknown; now in many places it usurps the ground, and especially overruns the districts which have been accustomed to the growth of wheat. But it is a punishment which has followed the practice of the ignorant and slovenly farmer, who has paid little attention either to cleaning his land at all, or to the right way of doing it, and who has continued for a series of years to take successive crops of wheat from the same exhausted fields.

The peculiarity of this weed consists in the hard covering with which its seed or nut is covered; in the time at which it comes up and ripens its seed; and in the superficial way in which its roots spread. The hardness of its covering is such that "neither the gizzard of a fowl nor the stomach of an ox can destroy it," and that it will lie for years in the ground without perishing, till the opportunity of germinating occurs. It grows up very little in spring, but it shoots up and ripens in autumn, and its roots spread through the surface-soil only, and exhaust the food by which the young wheat should be nourished. A knowledge of these facts teach — First, that unless care be taken to exclude the seed from the farm, it will remain a troublesome weed for many years, even to an industrious, careful, and intelligent cultivator. In the second place, that spring ploughing will do little good in the way of extirpating it, as at that season it has scarcely begun to grow. Thirdly, that raising wheat year after year allows it to grow and ripen with the wheat, and to seed the ground more thickly every successive crop. It is said that, when it has once got into the land, two or three successive crops of wheat will give the pigeon-weed entire possession of the soil. It is not, therefore, the immediately exhausting effects of the successive crops of corn alone which has almost banished the wheat-culture from a large part of North America, where this grain used to be produced in great abundance, but the indirect or after consequences of such a mode of culture in a considerable degree also.

In a previous chapter, when describing the geological structure and attendant agricultural character of western New York, I have described the Marcellus shales as forming the surface of the upland country, immediately behind the wheat-region properly so called. These dark-coloured shales form stiff soils, which produce good grass, but are difficult—naturally, or have become so by unskilful treatment — to retain profitably in arable culture. In Yates County, upon this formation, the pigeon-weed has become in some places almost the lord of the soil. It was unknown there, as elsewhere, thirty years ago; now "hundreds of bushels of the seed are purchased at the Yates County oil-mill; and if it were worth 8s., instead of 1s. 6d. a bushel, these hundreds would be thousands!"

The reader will observe, in the concluding words of this quotation, how one evil leads to another. The purchase of this seed at the oil-mills can only be for the purpose of adulteration. I have examined samples of American linseed-cake in which seeds were to be recognised which I could not name. They might, I then supposed, be those of the dodder, a parasite which infests the flax plant in some localities; but they might also be other cheap seeds purposely mixed with the linseed. Those who are in the habit of buying cheap American cake may think this point deserving of their attention. As oil-cakes are chiefly bought by farmers, perhaps it is only a kind of retributive justice that a set of idle farmers in one country should thus be the means of punishing a less discerning set in another.

One purpose for which I have dwelt upon this weed is to show that a knowledge of the habits, or physiological history, of our common plants is as necessary to the improvement of the art of culture, of the condition of those who practise it, and of the agricultural productiveness of a country, as almost any other kind of knowledge. No one will readily accuse me of a wish to undervalue the usefulness of chemistry to agriculture, and yet I have often had occasion to regret the evil influence of opinions hastily expressed by ill-informed persons—as if this were the only branch of knowledge which was necessary to bring this most important art to speedy perfection. It may be partly conceit, but it is chiefly ignorance, which has led many young persons, slightly acquainted with chemistry, to propagate crude notions as to the omnipotence of chemistry, and of the researches of the laboratory, in determining all difficult, doubtful, or disputed questions in practical agriculture. The longer a cautious and *safe* man lives, the less will he value extemporaneous opinions on matters which fall within the range of what may properly be called scientific agriculture—and the wider will appear the range of knowledge, theoretical and practical, which is necessary to the accurate solution, even of what some look upon as simple and superficial questions.

Among the crops which, in Lower Canada, have taken the place of the formerly abundant wheat, the oat is one of the most important. I did not learn what weight they here average per bushel, but the price, during the last year, has never exceeded 14d. currency per bushel. The Canadian farmer is content to sell his oats for this price, and to buy a barrel of pork with the proceeds, rather than expend his surplus grain and time in feeding his own pig for his own family, and making manure for his farm at the same time.

I have spoken of the long-limbed pigs of the French settlers in Madawaska. The pigs of this district, like their masters, are full cousins-german to those which inhabit the Upper St John. The snout and the legs, as they say here, *run a race for length*. The native cattle are also poor, and the starvation-limit system of feeding, generally practised a century ago in Scotland, is still in full force in colder Canada. It is to be hoped that the cattle-shows, which are now promoted by local societies, and encouraged by legislative grants, may be the means of gradually introducing a better system of feeding, with more profitable breeds of stock.

To obtain a view of the country, I climbed the hill of Belœil. On the top is a cross and chapel, and it is a place of popular pilgrimage. It is a pleasant, but a steep and laborious climb. The ascent is divided into fourteen parts or stations, at each of which is, or was, a cross, bearing an inscription having reference to the journey of our Saviour, as he bore the cross to the place of crucifixion. Pious devotees, who visit the mountain in considerable numbers, rest a while at each —as at the similar stations on the Rigi and other places of pilgrimage in Roman Catholic Switzerland—and, while they rest, repeat the appropriate prayers. These are printed and sold under the title of "Meditations and prayers adapted to the stations of the Holy Way of the Cross." Without underrating the efficacy of these devotions, the site of the chapel and place of pilgrimage are well selected. The healthful exercise, the free air, and the lovely view from the summit of the mountain, must lighten the heart, and send home the most sorrowful and careworn with a more cheerful spirit.

Looking towards Montreal, the River Richelieu flowed at my feet, far down in the valley—a long, scarcely sinuous, silver ribbon, singularly narrowing as it descends from its source in Lake Champlain on the south, to its confluence with the St Lawrence at Sorel, far away to the north. On its surface, a few specks showed where the passing craft were carrying the produce of Canada or the merchandise of Europe to their respective destinations. Villages and church-spires occurred at intervals above its banks; and far as the eye could reach, on either hand, a seemingly flat plain extended from the Richelieu to the St Lawrence, and beyond this—broken only by the height of Mont Royal—to the north-west, into the counties on the left bank of the great river, till the sky seemed to rest on the still-extending flat.

Turning towards the north-east, the same level land

stretched away on the other side of Belœil towards St Hyacinth; and, in the interval, flowing also towards the north, ran the smaller river Yamaska, pouring its waters into the St Lawrence in the middle of Lake St Peter, between the mouths of the Richelieu and the St Francis. Over this eastern part of the plain arose occasional isolated mountain elevations, which towards St Hyacinth in front, and in the direction of Lake Champlain on the right, appeared to thicken into clusters or ridges of loftier elevations.

This wide flat margin, of varying breadth, which on either side girdles the St Lawrence, accompanies it along a great part of its course, follows the Richelieu up to Lake Champlain, and thence stretches along its shores towards the city of Albany and the river Hudson. It is a post-tertiary region, the latest elevated part of north-eastern America, and contains the remains of marine animals, which are still living in the sea-mouths of the St Lawrence, and along the Atlantic borders of the New England States. In this part of Canada it rests on the Utica slates, and Lorraine shales of the New York geologists, through and among which, beds and domes of trap occasionally force their way, forming the isolated hills of which I have spoken. This Canadian plain exhibits, for the most part, a surface-soil of a pale yellowish-coloured clay, more or less heavy. Along Lake Champlain, however, and at higher levels in the States of New York and Vermont, and towards the Canadian border, the subjacent clay-deposit is covered with drifted sand and gravel, of various depths, which give to the available soil a more open character, degenerating in some places — as between Albany and Schnectady, already described—into an almost worthless pine-bearing sand.

When first cleared, this post-tertiary belt of level clay land—the St Lawrence Flats, we may call them—produced rich harvests of wheat, and, over a very large portion of their surface, continued for many years to yield abundant crops at little comparative cost. Hence the banks of the St Lawrence were deservedly called, in former days, the granary of Canada. The lower province could then afford, therefore, to export much grain; and among lesser articles of production, linseed was one of which large quantities were at that time shipped to Europe. Now, the shipment of either of these articles, the produce of Lower Canada, may be said to have virtually ceased.

We need not inquire in Canada after any special causes for a change almost equally marked in every other part of north-eastern America which has been as long under the cultivation of European settlers. Everywhere idleness, ignorance, and an avaricious spirit, on the part of the cultivators, have led to the same results in diminishing the ability or disposition of the soil to produce good crops of wheat. To speak figuratively, the spirit of fertility is every year retiring farther towards the west, shrinking from the abusive contact of European industry, towards the head-waters of the Mississippi and the St Lawrence.

And yet the peculiar tenure of land in Lower Canada, if it be not necessary to account for the existing condition of the soil, may possibly both have aided in bringing it into that condition, and may stand in the way of an easy or rapid improvement or restoration of its productive capabilities.

Of the whole lands in Lower Canada, only eleven and and a half millions of acres have yet been disposed of; and of these, seven and a half millions are grants in *fief* and *seigneurie* by the crown of France. Of the remaining four millions of acres of granted lands, many of which are large grants—such as that of seven hundred thousand acres in the eastern counties to the British American

Land Company—only one and three quarter millions are held by resident proprietors; and, of these, only about six hundred thousand acres are in actual cultivation. By far the largest portion—the great bulk of the land, we may say — is held by the habitants under the seigneurs ; and this comprises those tracts of country which, in the early settlement, were considered the most desirable — in fact, all the old-settled and originally productive wheat-lands on the St Lawrence.

The rights reserved by the lords of the soil vary, I believe, in kind and amount; but a fine of one-twelfth* upon every sale or succession, the right of pre-emption by the seigneur in every case of sale, by payment of the price offered by the highest bidder, and the annual reserved rent, are the chief burdens to which the holder is subjected.

In the first instance, the seignorial tenure is an advantage to the farmer, and affords facilities for settling in life to the young man who is destitute of capital. He goes to the lord, obtains permission to occupy a portion of wild land which is measured off, on condition of paying a small annual rent; builds himself a hut with the help of his friends, and begins to clear and sow. With a little provision, and a few tools, he can thus begin the world with scarcely a penny of money in his pocket. So far the seignorial tenure is favourable to the poor Canadian, though it is unsuited to the wants

* It is interesting to observe how important a place the payment of a twelfth or a fourteenth used to occupy in the social arrangements of our forefathers in purely agricultural districts. The Seigneur here takes one-twelfth or one-fourteenth of the corn for grinding it at his mill, to which, as on many estates still in our own island, the tenantry are bound to take their corn, and he receives a twelfth of the whole value of the land at each death or sale (*tods et ventes.*) In Hungary, at the present moment, near the course of the Danube, the rural labourer receives one-fourteenth for cutting the corn, and again one-fourteenth for thrashing it.

and wishes of the emigrant who, coming with a small capital, wishes to buy the fee-simple of a farm he can thenceforth call his own.

The reserved rent varies considerably. In the old grants it is fixed at a mere nominal sum. The farms, as I have already remarked, are long and narrow. They are generally three lineal arpents in breadth, fronting the road, and forty, thirty, or twenty, in depth. For these lots, the reserved rent averages about twopence sterling an acre; but, in many recent concessions, it is as high as fivepence sterling, or sixpence currency, an acre. Added to the reserved twelfths, and other privileges of the seigneur, this latter rent-charge is by no means a light one in Canada. Such a money-charge, being fixed, would not with us, in most places, seriously affect agricultural improvement; nor would it prove a serious hindrance to the introduction of a better system of culture in the exhausted flats of the St Lawrence, were it not that the farmers are already very poor. This small annual payment, therefore, is felt as a great burden, and is a source of much and general complaint —though I have not heard that they have actually, anywhere in Lower Canada, risen in special rebellion against the payment of what must be considered to be just debts, as their more enlightened and energetic Protestant neighbours, the New Yorkers, lately did in regard to the Renselaer rents. It is important, as indicating the general feeling in regard to these reserved rights, even among the French population, that in the Rebellion of 1837, Dr Robert Nelson, in his manifesto to the Lower Canadians, declared "for independence, a republican government, the confiscation of the crown and church lands and the possessions of the Canada Company, *the abolition of seignorial rights*, and imprisonment for debt." This promised abolition was, no doubt, intended as a bribe to those who had services to render in payment of the lands they held; but it, nevertheless, indicates that the existence of seignorial rights is considered a grievance by the people, and that, like our former tithe-payments, though strictly just, they will stand in the way of the agricultural improvement of the country. Every encouragement, therefore, should be given to the buying up of the annual rent-charges, and enfranchisement should be made compulsory as regards the fines on transfers, the right of pre-emption, of milling the corn, and other minor claims of the seigneur.

Sherbrooke, on the River St Francis, and on the proposed continuation beyond St Hyacinth of the St Lawrence and Atlantic railway, is near the centre of the grant made to the British American Land Company, which I have already incidentally noticed. This company prefers to sell their lands at from 10s. to 15s. an acre, to give credit to the purchaser for ten years, on payment of 6 per cent interest on the purchase-money, and, at the close of this period, to receive the price in four yearly instalments, bearing interest also while unpaid, at the rate of 6 per cent per annum. This tedious and complicated mode of payment does not appear to have found much favour with purchasers or settlers, and, with the inland situation of the district, and the neighbourhood of a French population, has probably contributed to the very slow rate at which land has been disposed of in these eastern counties, and at which its market-value has increased.

In the Report published by the Board of Registration and Statistics at Montreal, in 1849, it is stated of the county of Sherbrooke, that "the registrar does not know of a single new settler having located himself in any of the new townships of the county; nor does he think there has been any increase in the value of lands for ten years." This is a very unpromising account of the district, in which a large part of the grant of the Land Company is situated, and does not at all encourage intending emigrants to settle there. Of the adjoining county of Shefford, in which also they own much land, it is said in the same report, that about 1000 acres of the land sold during the year has been to new settlers, at the rate of 15s. to 20s. currency an acre, and that the increase in the value of land, within ten years, had been 5s. or 6s. an acre.

These eastern counties of Lower Canada were at one time brought prominently before the British public, in consequence of the late Mr Galt having gone out to Canada as agent to the Land Company. But if we may judge from the proclamation issued by Dr Nelson, during the time of the Rebellion of 1837, such companies are not very popular with the French Canadians, nor settlers upon their lands regarded with much favour. And without supposing that there is anything wrong either in the existence or management of these companies, we can understand why the old settlers should look with disfavour upon bodies of men whose professed object is to bring in emigrants of another blood, tongue, and faith, and, at the same time, to lessen the facilities and enhance the expense of settling their own increasing families.

I regretted that my leisure did not permit me to visit Sherbrooke, for the purpose of inspecting the district in person. It is said, and I should think with much reason, that the Company's grant, along with the whole of the inland region of which it forms a part, will be greatly benefited by the completion of the St Lawrence and Atlantic railway now in progress.

Sept. 27.—Unable longer to avail myself of the kind hospitality of Major Campbell, I left St Hilaire at 8 A.M., and returned by railway to Montreal, along with Mr Wettenhall. I would strongly recommend flying visitors to Montreal to devote a day to an excursion to St Hilaire, and a climb to the top of Belœil. They will be able to procure agreeable accommodation at an hotel which the seigneur was building at the time of my visit, in a beautiful situation, for the accommodation of railway tourists.

In the afternoon I visited, in company with some agricultural friends, the brickmaker in whose hands the tile-machine of Major Campbell had been placed. He had this season made 40,000 tiles, all of which he expected to sell at the price of six to eight dollars a thousand, according to the size. It was chiefly a few British farmers in the neighbourhood of Montreal who had hitherto tried them, but so far with much advantage. I am satisfied that, in all the St Lawrence flats, they are to be a means of much agricultural improvement. As they become more in demand, their price at the tile-work will diminish, and the cost of executing thorough-drainage be in consequence lessened.

The plea will not be so generally urged here as it is in the New England States, and in that of New York, against the expenditure of money in improvement, "that the land, when drained, will not sell for an equivalently increased price in the market." For, though it may be equally true here as in the States, yet the French Canadians are a more fixed, home-loving race of people, not so given to change, and would therefore, if they had the money, be more willing to expend it in improving and embellishing the houses of themselves and their children. But the vast number of mortgages with which the farmers in Lower Canada are oppressed may prove an obstacle, which only a board of "Commissioners for the Sale of Encumbered Estates" will be able to overcome.

I had only time left to take a hasty drive up and around the hill of Montreal—an excursion to which I would gladly have devoted an entire day. The rocky surface of the island consists, for the most part, of the same Trenton limestone on which Kingston stands; but it is interstratified with greenstone trap, of which an outburst forms the Mont Royal. Independent of the drift, which deeply covers the hollows and slopes of the island, and modifies its natural surface, there are in the mingled *debris* of these two rocks materials enough to account for the fertility which in ancient times made it the central residence of the Indian tribes, and has since secured it the frequent eulogies of French and other writers.

At half-past six P.M., I went on board the steamboat for Quebec. The weather was thick, dark, and rainy, and as I knew no one on board, I retired to my stateroom at seven. After a rainy night-voyage of a hundred and sixty miles, during which nothing of the country on either side of the river was to be seen, I found myself in twelve hours in a quiet pleasant room in the St George's Hotel at Quebec.

In leaving Montreal, it was a matter of much regret to me that I had been obliged to forego the pleasure of making a tour up the Ottowa—a river which is inferior in size only to the St Lawrence, which runs through a country interesting in very many respects, and is the natural outlet for the drainage of an area of eighty thousand square miles. The vast region to which this river and its branches afford the means of internal navigation, and of communication with external markets, is already extensively settled. It has also a rapidly increasing commercial capital of 12,000 inhabitants at Bytown, where the Rideau Canal—150 miles in length—leaves the Ottowa for Kingston, on Lake Ontario. When the first period of settlement has passed over, during which the lumber-trade occupies the sole attention of nearly all the settlers, and the population shall have become mainly agricultural, this district will assume a permanent importance, in reference to Canadian strength and resources, which will every year rise in public estimation.

I have alluded but very slightly to the political differences and excitement of which Montreal was still, to some degree, the scene at the period of my visit. I had as yet enjoyed too few opportunities of acquiring a knowledge of existing local and provincial circumstances, feelings, and prejudices, and of their antecedents, to enable me to form a satisfactory opinion as to the right and the wrong of all that had been done.

In judging of public events that fall out in a place like Montreal, however, much allowance ought to be made for the peculiarly heterogeneous character of the population, and the frequent opposition of their interests. Thus the population of 50,000, which the city is said to contain, consists approximately of—

French Canadians, ...	20,000
British Canadians, ...	10,000
British and Irish born, . . .	15,000
Germans—United States born, &c. .	5,000
	50,000

Out of these different nationalities arise, as separate and opposing parties—

1. Those of British, or Anglo-Saxon, against those of French blood.

2. The Canadians, born of British blood—the United Empire Loyalists, &c.—against the home-born, or native British. The seat of these parties is properly in Upper Canada; but Montreal contains a large section belonging to each.

3. Among the French—the party of Papineau, opposed to British connection, against that of Lafontaine, which at present is in favour of such connection.

4. Then partly interfering and partly coinciding with these are the Radical and Conservative political parties.

5. The religious parties—the Protestant against the Roman Catholic.

6. Also in Montreal, as a seat of commerce, there are the agricultural and commercial parties; among whom the free-trade movement, as with us, but especially in its relations to timber, and to reciprocity with the States, has been a great bone of contention. It is the latter of these parties which has mainly supported the Annexation movement.

7. And, lastly, city and municipal affairs have formed parties purely local, whose feelings on these matters complicate the other differences.

In these numerous parties, among a small population of 50,000, just large enough to furnish materials for giving a certain degree of importance and consistency to each, without affording a marked preponderance to any, there are surely abundant materials for easy disorder and sudden excitement. And if it be farther the case, as I was generally informed—correctly, as I should judge from my own limited experience—that in Montreal all these parties are very bitter against each other on occasions of excitement, and, even in ordinary times, co-operate little with each other, the wonder will be rather that the city should have remained so long quiet, than that a single boiling over should have satisfied a perpetually simmering population.

It must be a very angelic Government, indeed, that could please so divided a people, a very talented one that could persuade, and a very powerful one that could restrain or control them. Military men, like other leeches, are naturally partial to their own mode of cure, and blame the Government for want of energy and decision in not employing the strong arm to repress them during the late disturbances. But people at home will scarcely condemn a governor for preferring to attain his ends by peaceful means, rather than by force of bloodshed. So far as I have had the materials for forming a judgment, it appears to me to have been not less wise and prudent than humane to quit the city altogether, and to leave it to make up its own intra-mural differences. A punishment which must affect their pockets will be far more felt than one which might have robbed the city of a few of its most worthless lives, and will sooner work its way to the understandings of its inhabitants.

There was one point, however, in regard to the Canadian differences, which, from my previous ignorance of the province, was incomprehensible to me, and upon which I anxiously sought to be enlightened. When the rebellion broke out among the French Canadians, the Upper Canadians, and generally those of British blood, took part against them. But in the division in the House of Assembly upon the Lower Canada Rebellion Losses Bill—the cause of all the disturbance—a majority of the Upper Canadian members, of British blood, and many of them British-born, voted with the French members of Lower Canada in favour of the measure. It was not, therefore, a war of races, as it had been represented at home. But how came those who were unanimous in opposing the rebellion to be found voting with those who had favoured, or actually supported and participated in it? I put this question to a friend of mine, one of the Upper Canada members, himself British-born, who had voted for the obnoxious measure, and his explanation was to the following effect:—

"For a long series of years, Upper Canada was under the dominating rule of what was called the Family Compact. Home-born Canadians, and a certain number of high officials, divided all offices and patronage among themselves, and did everything in their power to keep the British-born from participating in the exercise of influence, or in the sweets of office. The few British who gained access to the Assembly, therefore, were naturally driven into opposition, and, after the union of the provinces, made common cause with the French opposition to the Tory Government. By degrees, however, the British-born in Upper Canada increased in strength, till at length the members of Assembly returned by them exceeded those nominated and returned by the Family Compact. They were then able, by the aid of the opposition members of French blood, to drive their enemies from office, and bring in that Government which now holds the reins of power. It was no way surprising, then, that a majority of British-born should be found fighting side by side, when in office, against the same parties whom they had joined to oppose before the reins of Government were intrusted to their hands; or that our ousted opponents should be bitter, and say all manner of evil against us.

"And then, as to this disputed measure, we never believed or intended that any one who had aided or promoted the rebellion should be compensated for the losses he had sustained; though some of our supporters spoke foolishly, which we could not help, and our leaders were not perfect by any means in their behaviour on some occasions. How, then, could we abandon our old friends? We felt, indeed, that we had no cause; and if we had found a cause, we could not, amid the clamour that was raised, have honourably taken advantage of it."

This defence places the question in a light I was too ignorant of the circumstances to have been able to see it in before. I could not reply to it; and as I was only asking for information myself, I place it before my readers, who may possibly be in the same state of happy ignorance with myself, without committing myself either for or against the statement of my Canadian friend.

CHAPTER XII.

Land opposite Quebec.—Its quality and value.—Reserved rents considered oppressive.—Few immigrants into this region.—Roman Catholic seminary at Quebec.—Professor Horan.—Self-sacrifice of the teachers.—Falls of Montmorenci.—The natural steps.—Ice cone of Montmorenci.—Sun-setting on Quebec.—Relative proportions of the different sects in Quebec.—Comparative commercial prosperity of Montreal and Quebec.—Autumn around Quebec.—Fires in the city.—Journey down the St Lawrence.—Price of land and labour at St Michel.—Flat lands of St Thomas, "the granary of the lower district."—St Roque des Annais.—Long farming streets.—Upper Bay of Kamouraska.—Mode of drying grain.—Price of farms.—College of St Anne.—Rapid increase of the French Canadian population.—Early marriages of the French population.—Healthiness of the climate.—Comparative births and deaths in Lower Canada and in England.—Corn-mills.—Kamouraska.—Rivière-du-Loup.—Comfortable hotel.—Village of Du Loup, and its future prospects.—High-road to New Brunswick.—Active Canadian horses.—Cacona.—Country apparently thickly peopled.—Great extent of wild forest-land in these lower counties.—Large families of the Canadian peasantry.—Subdivision of farms.—Poor and difficult land on which they settle.—Resemblance of the poorer habitants to the poorer Irish.—Desire to build fine houses.—Extensive mortgages.—Wages of labour in the Rimouski district.—Longitudinal valleys parallel with the St Lawrence.—Peculiarity of the bog-earth in North America.—Difficulty in finding quarters at Rimouski.—Irish landlord.—Scottish settlers at Mitis.

Friday, September 28.—After breakfast the weather improved, the sun shone brightly through the clouds, and I was able to walk out into the strong city and citadel of Quebec, where every gun and every soldier calls up the memory of the immortal Wolfe. The place is certainly very strong by nature; and no art seems to have been spared to turn to account the advantages of natural position. The only weakness, perhaps, is that the fortifications are too extensive; and in the event of a war, would require a larger force to maintain or defend them than could easily be spared in so extensive a country.

Not having met with any of the persons to whom I had brought letters, I crossed the river to Point Levi in the afternoon; and climbing the lofty bank, from which the view of the city and river is very extensive and beautiful, I made a short excursion on foot into the interior.

The rocks, which on this right bank rise up almost precipitously from the river—like the high ground and cliffs on which the city of Quebec stands—consist of dark-coloured slates or indurated shales, having thin beds of limestone, more or less pure, interstratified with them. They belong to the higher beds of the upper Silurian—the Utica and Lorraine shales which overlie the Trenton limestone of Kingston and Montreal—and are inclined at a very high angle. From the elevated ground beyond the top of the bank, the country inland appeared to be cleared to a great extent, and undulated in long wave-like ridges, till the eye finally rested on low mountains, which I supposed to be a prolongation of the Green Mountains of New Hampshire and Vermont.

The soil was free, and comparatively light, being formed for the most part from the crumbling of the shaly rock, of which many fragments were intermingled with it. Indeed in some fields, where the rocks protruded at intervals through the surface, the soil—like that fertile country of which I have already spoken, near Woodstock in New Brunswick, or the still richer fields on the Onondaga green shales near Syracuse—consisted almost entirely of visible fragments of the shivery indurated shales.

Potatoes were almost the only crop I saw now remaining in the fields. On his knees, working with a diminutive hoe, I found a French Canadian digging, from what appeared a good soil, a scanty crop of small-sized potatoes. The man seemed satisfied, and considered the crop good—perhaps it was, in comparison with those of the past years of failure. Their size reminded me of the graphic if not elegant terms in which a New Brunswick farmer, on his return from Maine, described the potato-crop of that State to a friend of mine, in the autumn of 1849: "Why, sir, the potatoes in Maine—I could put fourteen of them into your mouth, and sixteen into that of any other man. I could, by thunder!—I have said it!"

The peasantry obtain land from the farmers and proprietors for their potato-crop—as is still the case in many rural districts among ourselves—free from rent, on condition of manuring and cleaning it. The owner takes a crop of grain the following year, and again lets it for potatoes on the same terms, if he can. Farther inland, my informant said the land was let to tenants at so high a rate that they were much kept down by it. On further inquiry, he stated the rent to be a piastre the arpent*—which meant a dollar for an arpent in front, and perhaps thirty backwards—which is about twopence an acre. What can be the condition of an agricultural people, who, paying scarcely any taxes, feel themselves weighed down by a rent of twopence an acre?

I was here in the county of Dorchester, and though it was situated directly opposite to Quebec, there had not come into the county a single new settler during the year 1848. The value of cleared land varies from £1 to £10 currency. There are two reasons for the custom, prevalent in all Lower Canada, of running the farms back from the roads, with a narrow frontage. Society is

* Six arpents make about five imperial acres.

obtained, so necessary to the French, and many neighbours close at hand, by living in one long street. Many farms can also be laid out, with little expense in road-making. Behind the first row of farms along the road, a second row is surveyed; and beyond these a second road, and a third row of farms. These are spoken of as the second and third Concessions. They are farther from churches, markets, mills, &c., than the first Concession, and are therefore less esteemed. In this county, at present, the seigneurs grant the uncleared back-lands for an annual charge of 15s. to 30s. currency for the farm of 90 arpents—3 in front, and 30 deep—or 2d. to 4d. an acre. This is considerably greater than it used to be, as in this county the value of uncleared land has risen much during the last ten years.

Immediately above this county, along the St Lawrence, is that of Lotbinière. Here cleared land sells for 50s. an acre, and uncleared land has not increased in value for ten years. Immediately below, again, is the county of Bellechasse, in which also there are no new settlers. Cleared farms, well situated, will sell for £400 to £600; but the general value of land during the last ten years has fallen 10 or 12 per cent.

Inland again, beyond and behind Dorchester, is the county of Megantic, which is described as very flourishing. There are no immigrants; but the population is rapidly augmenting by natural increase. It is a purely French Canadian district. The average value of cleared land is about 6s. 3d. an acre, though some sells as high as 20s. It produces sheep, cattle, and butter for the Quebec market. It will be recollected that the price of cleared land is not that of the fee-simple, the rents and rights of the seigneur being always reserved. Probably, this form of holding is one of the causes by which immigrants are deterred from penetrating into these Lower Canadian counties.

Sept. 29.—Among other places, I this morning visited the Quebec Seminary, a Roman Catholic institution, founded nearly two hundred years ago, which accommodates and instructs 180 boarders, and 120 day-scholars—a number about equal to those taught in the seminary at Montreal. The boarders pay £17, 10s. currency per annum, and the day-scholars 20s. I was much interested in the Professor of Natural History and Chemistry, Mr Horan, who had visited and studied at the universities of Cambridge in Massachusetts, and Yale in Connecticut, with the view of obtaining instruction which was beyond his reach in Canada. I found him and others fully alive to the importance of an improved agriculture in Lower Canada, and anxious to aid in introducing instruction in its principles into the elementary schools of the lower province.

I was never so much struck as upon this visit, and by my conversations with Mr Horan, with the meritorious and self-denying spirit of the teachers in some, at least, of these Roman Catholic seminaries. This institution has a small endowment, but is not rich like that of Montreal, and barely makes ends meet. The professors, without any prospect of ever rising to any position beyond the walls of the seminary, devote their lives to the duty of teaching without remuneration. They are lodged, fed, and clothed, and at vacation-time, if they choose to visit anywhere, they are allowed twenty dollars for expenses. Thus they live and labour from year to year, with scarcely any society beyond the walls of the institution, shut out from the cordial of human sympathy, and centring their affections on a brighter future. I do not praise the system, regard it as natural, or consider it the way in which talent may be best employed for the benefit of our fellow-creatures; but for the self-denying spirit of the devoted clerical teachers, I could not help feeling a sort of reverence. There were men of fine minds, fresh affections, and warm feelings, capable of attaining worldly distinction in their several walks, of enjoying the intercourse of social life, of being happy in the exercise of domestic affections—laying down all worldly hopes and prospects, and sacrificing themselves to the duty of instruction. While I pitied, I could not but respect the men—feel there was something in them higher and nobler than had moved myself in my struggles through life; and while I almost felt indignant at the inhumanity of a system which could exact it, my heart warmed the more towards my friend Horan, who had been able voluntarily to sacrifice himself to it.

I know there are many mindless and heartless beings, male and female, to whom such a devotion would prove no sacrifice. I do not allude to such persons, nor to the dry and unfeeling mummies into which the lapse of long years of routine may convert even those who at first made a true and priceful sacrifice. But I saw before me men bright with intelligence, and with a capacity for appreciating enjoyment still unseared; and I could not but honour them for their self-sacrifice, because I knew them to be still human enough to feel that it was great.

In the afternoon, I drove nine miles down the left bank of the St Lawrence, to visit the Falls of Montmorenci. Though the quantity of water which descends is insignificant after Niagara, and smaller even than at the Grand Falls on the St John River, yet it descends from a height of nearly 250 feet, and the place is well deserving of a visit. The edges of the highly-inclined slate rocks are overlaid by nearly horizontal beds of impure thin limestones, which are cut through by the river, and eaten back for several hundred yards from the St Lawrence, till a hard metamorphic gneissoid rock has arrested the cutting process. Over this the water at present falls. Upon the horizontal beds is a deposit of 10 to 30 feet of drift; and upon this, adjoining the St Lawrence, at a level somewhat lower than the top of the Falls, a deposit of 1 to 6 feet of yellow marine sand, mixed with recent shells.

A mile above the Falls, on the same river, occur what are called the "natural steps," where the horizontal beds of comparatively soft rock are cut by the water into deep ravines or gullies of a very romantic character, and in many places form series of natural steps, from which the place derives its name.

The most peculiar circumstance in connection with the Falls of Montmorenci is the appearance presented on the channel of the river a short distance below the cascade, when winter sets in. When the stream below becomes covered with ice, the falling spray descends and collects upon its surface in showers of snow, which cohere and harden, and gradually accumulate into a lofty cone of ice—having the living cataract behind, and the broad, still, frozen plain of the St Lawrence in front.

This conical hill forms a natural "Montagne Russe," much frequented by the young people of Quebec, being a convenient distance for sleighing parties even in the shortest days of winter.

In the immediate neighbourhood of Quebec, land sells high. For a little farm of 70 acres, on which his house stands, immediately above Wolffe's Cove, Mr John Gilmour told me he had paid at the rate of £75 currency an acre. Near all thriving towns, in the New World as in the Old, land, even for farming purposes, brings a comparatively high price. Near this city the land is very good in many cases, and generally produces excellent green crops. In both Canadas, as now in Ireland, such crops are becoming more cultivated since the potato became less certain. Mr Sheppard, the well-known seedsman in Montreal, informed me he had this season sold twice as much turnip-seed in Lower Canada, and twenty times as much as usual in Upper Canada. Of mangold-wurzel seed, four times as much as usual had been purchased by the Lower Canadians, as experience had shown it to be better adapted to the climate. This may be because of the extreme heat of the summers on the river St Lawrence.

Formed from softish, somewhat calcareous slates, which in many places are near the surface and crumble readily, the soil is inclined to be heavy, and rests often on an impervious bottom. Drainage, therefore, generally, and the use of lime in many places, are indicated as means of improvement. The latter, if I may judge by the frequent lime-kilns I passed on my way to Montmorenci, is tried to some extent by the farmers around Quebec.

New settlers seldom remain in this county. The average value of uncleared land is about 5s., and of cleared from 15s. to 17s. 6d. an acre. During the last ten years, the value of land has increased one-fourth.

On my way back from Montmorenci, about six in the evening, the quickly setting sun shone on the tinned roofs and spires and glittering windows of Quebec, producing an ever-varying but very beautiful effect. I had in my life before seen only two sunsets more striking, and of which it reminded me—that of Paris from Montmartre, on a clear autumnal evening, when the sinking rays of the sun lingered still on the Panthée, on Notre Dame, and other prominent objects, singling them out as individual pictures from the countless mass of objects of less elevation; and that of Moscow from the Sparrow Hills, when its thousand domes of gold and silver, with intermingled green and azure, and its ten thousand ornamental crosses, glittered around the ancient Kremlin, and the massive central palace-fortress itself rose as a huge dark pile, frowning from its Eastern towers on the congregated churches and other buildings of the city beneath.

Sunday, Sept. 30.—After a brief visit to the Roman Catholic cathedral, where I saw the venerable, white-haired, old Archbishop of Quebec, the Primate of the

North American provinces—since dead, I believe—assisting at early service, in the midst of a large and apparently devout congregation, I went to the Scotch Presbyterian church, where I heard an excellent sermon. In the churches already in use, and in the appearance of the new ones in course of erection, there are no signs either of pecuniary depression or of a want of general zeal in matters of religion.

The relative numbers of the several religious sects in Quebec is considerably different from what it is in Montreal. In 1844, the last published census, the Roman Catholics formed seven-ninths of the whole population, then 46,000, now probably upwards of 50,000. Those of the Church of England formed only one-ninth, and of the Church of Scotland one-eighteenth—all the other sects made up the remaining eighteenth. It is, perhaps, because the predominance of the Roman Catholic French Canadian population is so great that party differences, whether political or religious, are represented as being much less bitter here than at Montreal.

If I were to judge from my own experience only, I should say that political differences of opinion, in reference to recent events, were at least as bitter as they could be at Montreal. During my short stay in Quebec, I met at dinner a native of Great Britain, but an old resident and a prosperous merchant in that city, who, after discussing the Rebellion Losses Bill and the Governor's conduct respecting it, hastily wound up his observations by observing, "that he would himself have helped to tar-and-feather him." I laughed, and said it amazed me much to hear such a person as he talk so violently. He evidently had not meant what he said, and remarked, "Well, I have never spoken so violently before; on the contrary, I am considered too moderate, and am obliged to keep arms in my house to defend myself, in apprehension of an attack from the violent people, because of my moderation." "If the moderate people of Quebec talk as you do," I said, "I wonder what your violent men can say."

It is feeling, and not judgment, that has led to the excitement and extreme language and action of these British Canadians. I could not help remarking to myself, in reference to the Quebec merchants, that if, in their ordinary affairs, these successful and prudent men were to show a tithe of the want of consideration displayed in their political conversation, neither their own nor other people's business would so prosper in their hands.

But I feel that I should be wrong to judge of the opinions of the people of Quebec from my own limited experience, or to estimate the sentiments even of a single British resident from a hasty after-dinner expression. There are at present reasons of an economical kind why the mercantile community in Montreal should be in a more excitable condition than that of Quebec, and which reasons may possibly go far to explain the angry freaks of the population of the former city. From some cause or other, which it is not necessary here to investigate, the trade of Montreal had, for several years before the late *émeute*, been seriously declining, while that of Quebec had been largely increasing. This will be seen by the following comparison of the imports into the two places in pounds sterling:—

	Montreal.	Quebec.
1841, . .	£1,699,837	£179,109
1845, . .	2,153,631	585,533
1847, . .	1,695,978	655,000
1848, . .	1,217,604	514,393

I have inserted only as many years as were necessary to show that the imports of Quebec have greatly increased during the last ten years, while those of Montreal, notwithstanding the rapid growth of the province, have rather diminished. The commerce of the former city is in a more flourishing condition, therefore, than that of the latter. And though, in the dull year of 1848, the imports into Quebec fell off considerably, it was only for that one year, while, from 1845, those of Montreal had been rapidly lessening. In the exports, also, from the two places, there was a much greater comparative falling off in Montreal than in Quebec during the year 1848.

Every observer of the political atmosphere knows well that, under such circumstances, governments, whether home or provincial, are always thought to display less ability, and their measures to be much less suited to the wants of the times and to the regard which ought ever to be had to the rights and liberties of the subject, than they are generally esteemed to be when harvests are good and commerce flourishes.

Autumn had now fairly set in in the province of Quebec. About Kingston and Montreal, and upon the banks of the St Lawrence, as I descended the river, the leaves were still green upon the trees, the weather generally mild, and the mid-day sun hot and powerful. But the beautiful amphitheatre of mountains towards the north, and east, which gives one of its peculiar features to the striking scenery around Quebec, had this morning clothed itself in the bright autumnal tints—of which I have already spoken as characterising the woods of North America in the *fall* of the year. Like our Highland hills in summer, when the heather is in bloom, the mountain ridges appeared more cheerful and lively for the change, and their outlines and varying features were more distinctly brought out to the eye than when the universal green covered with a common hue both mountain and plain, and blended together narrow ravine, wide retreating hollow, rocky escarpment, and rounded outlying hill, in the hard and less distinguishable outline of a lofty, dark-coloured, and frowning rampart. The sharp air, too, indicated that the season for summer clothing was nearly over, and that blanket-coats and homespun cloth, and even buffalo-robes and dresses, must be speedily provided.

All Europe is aware of the superior ease and expedition with which the people of Quebec contrive to enlighten and enliven their wooden city with frequent fires. Three successive times the fire-bell had already rung since my arrival; but the affairs had been small, all of them during the day, and had attracted little general attention. But things were differently managed to-night. As I was falling into my first dose, the bells began to sing out, a running of many feet followed, and then the clattering of engines, and by-and-by a light began to fill my bedroom; and as I lay in bed, the playing of a huge flame became gradually visible, rising and falling above the houses on the opposite side of the square in which the hotel was situated. All night long, the fire and the light, and the varying flame, and the noise of the bells, and the mingling tones of many sounds and voices, continued, and when daylight returned, four houses were already consumed. But the flames were subdued and danger over; and as I emerged into the street to prepare for my departure from the city, the weary firemen and their soiled engines were again on their way homewards.

Monday, Oct. 1.—After an early breakfast, I crossed the river to Point Levi, and soon after nine A.M. was on my way down the right bank of the St Lawrence in a light caleche, guided by a French Canadian, and drawn by a single horse. I proposed to descend the river as far as the Mitis, and thence to cross the province of Gaspé, by what is called the Kempt Road, to the Restigouche River in New Brunswick. All this I proposed to accomplish in the course of the week, and to arrive at the town of Campbelton on the Restigouche on the evening of Saturday.

Among the more striking objects of this day's journey was the Fall of Montmorenci, which I had visited on

Saturday. During the latter half of the first stage, as I descended towards the Isle d'Orleans, the mouth of this river gradually opened on the other side of the St Lawrence, and the morning sun, shining on the broad white sheet of its falling waters, made it, in some respects, a more striking and impressive feature in the landscape, than when seen, with its accompanying roar and mist, from the immediately adjoining rocks over which it is precipitated.

By half-past two, I had reached St Thomas — a distance of thirty miles. The land, in general, was good; for the most part lig t, and easily worked, but sometimes of a stiff clay. As I passed through the parish of St Michel, eight or ten miles below Point Levi, I was informed that the crops it yielded were somewhat less than two hundred bushels of potatoes, or a ton of hay, an acre, and that it sold from four to four and a half dollars. It is let sometimes on halves, or for a money rent of eight shillings an arpent. Labour costs one shilling and eightpence to two shillings currency a-day in summer; but in the short days of winter not more than one shilling a-day. Men are hired by the year at eight to ten pounds currency, with board and lodging. Labour, therefore, is cheaper here than in either Upper Canada or New Brunswick—perhaps the services of French Canadians are in reality not so valuable.

In the back or more inland country, where the land belongs to the province, new Concessions are being established by the Government, in which sixty-acre farms —two acres in front and thirty backwards—are offered in fee-simple for thirty dollars, or at the rate of two shillings sterling an acre. There are no new settlers in these counties; but the native proprietors, as my informant stated, readily save money enough to establish their sons on such new farms. Where a man has several sons, he is allowed to select adjoining grants for each of them. Old cleared well-situated farms sell at four hundred to six hundred pounds currency in this county of Bellechasse, though of late years they have, both in this and in the adjoining county of l'Islet, considerably lessened in value.

At St Thomas—where the South River empties itself into the larger river St Francis, before their united waters descend to the St Lawrence—a large breadth of flat and apparently rich country intervenes between the mountains and the great river. At St Thomas, there are falls on the St Francis, upon which a Scotch settler has erected mills, which grind up the produce of this rich flat, formerly called the "granary of the lower district." Its produce is less now than it used to be; but the land is still very capable of yielding remunerative crops.

Skirting the St Lawrence still, I drove on through the county of l'Islet to St Jean du Port Joli, through which I passed as the evening was coming on. It is a cheerful clean little village, with many nice-looking two-story houses, of considerable size and some architectural pretensions; and several new houses were building, which must be regarded as tokens of prosperity. At nine P.M. I stopped for the night at St Roque des Annais, having come altogether only sixty-three miles. The last twenty miles I had travelled in the rain, and, for the most part, in the dark; it was, therefore, a greater disappointment to find I had got into bad quarters for the night. I would recommend those who may travel this road after me to endeavour so to arrange their route as to fix their night-quarters at some other spot than at St Roque des Annais.

As long as daylight continued, I was still travelling in the county of l'Islet; and although, as I have already said, the value of land, both in this county and in that of Bellechasse, has diminished of late years in consequence of the failure of the crops, yet I passed, during the day, in these counties, not only much fine land, and a thickly settled country, but what appeared to me, as a mere passer-by, to be many indications of prosperity and comparative comfort. About every two leagues, often nearer, rose a large church, and beside it a comfortable parsonage or *presbytère*, and houses enough to form a considerable village or church-town. Many small rivers flow into the St Lawrence in the county of l'Islet, which are navigable for short distances; and the new houses building in many places indicated that, notwithstanding the failure of crops, and the alleged scarcity of employment, all gleams of prosperity had not yet left this part of the lower province.

The stranger will here also see illustrated, what I had already become familiar with in Nova Scotia and New Brunswick, the peculiar love of society and neighbourhood by which the French population are distinguished. Continuous rows of houses, separated by one or two intervening fields, accompany him for miles along his road. In fact, wherever the country is fully settled, this is the case, unless the traveller happens to turn up a cross-road, when a couple of miles may occasionally be passed without meeting with a farmer's house. This close neighbourhood is obtained by the method already described, of dividing the land into long stripes, or narrow adjoining farms. It is much to be wished that those who really have the agricultural improvement of the French Canadian holdings at heart, should endeavour to have this system discontinued. The amount of labour, both for men and horses, is greatly increased by placing the centre of operations and the home of the labourers and stock at the extremity of these stripes of land; and the difficulty is greater in properly superintending and executing all the necessary operations of the farm. Separated more widely from each other, they might possibly gossip less and labour more.

Tuesday, *Oct.* 2.—Much rain had fallen during the night and the early morning, but it had cleared off before I started at half-past eight A.M. The fall of rain, however, was unfortunate for me, as it made the roads heavy as well as dirty, and delayed my advance very much.

Soon after leaving St Roque, I entered the county of Kamouraska. To this county, which is not densely peopled, a few new settlers have recently come, though I am not aware if any of them are from the British Isles. Uncleared land sells generally at half-a-crown an acre; cleared land as high as two pounds an acre, and, in some places, considerably higher.

Two bays distinguish the river coast of this county, called respectively the Upper and Lower Bays of Kamouraska. At St Anne la Pocatière the upper bay commences. It is encircled by a ridge of romantic rocky hills, against which, at some ancient period, the waters of the St Lawrence washed, but which is now some miles from the water's edge. The intervening space consists of a broad flat of rich land, in some places still marshy and wet, but where it is sufficiently elevated, bearing magnificent crops of hay and of oats, wheat and potatoes. The road runs through this flat for six or eight miles, and at the lower end of the bay crosses the river Ouelle by a bridge of considerable length. Through banks of rich alluvial matter this river here falls into the St Lawrence, and forms a small harbour, which is the shipping-place for the rich land of the valley.

Though marshy, I was informed that this flat is exceedingly healthy—as most places in Lower Canada and New Brunswick are said to be—even where in Great Britain fever and ague would inevitably prevail. But nevertheless, for agricultural reasons, it is a fit locality for the introduction of a general thorough-drainage. The narrow nine-foot ridges so common in Canada, the open furrows between them, and the large main-drains or ditches around the fields, are all insufficient to remove the water which falls and accumulates in the land. To keep the two sets of open ditches in order must here, as elsewhere, annually cost much more than the interest of the sums which the construction of covered drains would require.

Yesterday (1*st Oct.*) I had seen much hay in the act of being led, and to-day I passed many fields of oats still uncut, and a few of wheat, barley, and pease. Many of the fields of oats which were cut, but not carried—and such was the case upon this flat—were spread out in the fields, as in the grassing of flax, at the mercy of the wet weather, which appeared now to have set in. I did not see a single sheaf or stook during the whole day. This method of spreading out, instead of binding up the corn when cut, is nearly exploded, I am told, even in the higher parts of Canada East; but in these lower counties it appears to be still universal.

Five hundred pounds was stated to me to be the value of a farm on this land, two arpents wide by forty deep. This is £400 sterling for sixty-six acres, including house and farm-buildings—or at the rate of £6 sterling an acre. Along the coast from this point, indeed, this sum was frequently named to me as the value of improved farms, with good houses upon them; and it appears to be a kind of average price among the French Canadians of the Lower St Lawrence.

Towards the northern extremity of the ridge of hills which girdles this bay, and looking down upon the belt of rich land of which I have spoken, stands the college or seminary of St Anne. The building is large, handsome, and beautifully situated. It accommodates, and is occupied by, about a hundred and eighty students. The rate of board is £17, 10s., as at Quebec, and the whole expenses about £20 a-year. I was surprised to find in so remote a spot a college containing so many pupils, and interested in the fact that so many of the habitants should be in a condition to pay even so comparatively moderate a sum for the education of their children. This institution enjoys a considerable reputation, and has pupils from Nova Scotia and New Brunswick. There are twenty such seminaries, it is said, in Lower Canada; and it was formerly objected, as a matter of reproach to the habitants, that too many of their young men received at these numerous seminaries a classical education, unfitting them for ordinary rural labour, while too few received that elementary education which our common schools are intended to impart. This reproach is still in some degree applicable; but the influence of the United States and of Upper Canada must by degrees bring the school-system of the Lower St Lawrence more into conformity with the general tendencies of the age in which we live.

Near Kamouraska church and faubourg, the lower bay of this name also contains good flat land, and, with occasional interruptions, similar land stretches along the shore as we descend. Inland valleys also, parallel to the St Lawrence, exist behind the rocky ridges which successively occur as we advance into the interior, in which good land, more or less granted and settled, prevails. Should foreign settlers continue to shun this region, there is ample room for the natural expansion of the native population for many years to come.

And yet the French habitants in this cold region of North America increase very rapidly. For this increase there are several reasons. One is, the early marriages to which both sexes are addicted among the people of French descent. At Kamouraska I had stopped a few moments to obtain a fresh horse and carriage, and, on starting, expressed to my new *cocher* my admiration of his pretty young wife, and inquired her age. "One-and-twenty." "And how long have you been married?" "Six years; and she was a widow when I married her!" Fourteen and fifteen is a common age for the marriage of females, and eighteen for males, on the shores of the St Lawrence. These early marriages are the more likely to accelerate the increase of the French Canadian population, as the females retain their vigour, and continue prolific, to a comparatively advanced period of life.

Another cause is the healthiness of the climate. I have already adverted to the general opinion upon this subject, in regard even to localities such as in Great Britain are usually unhealthy. But the statistical returns for Lower Canada, in relation to births and deaths, bears out this general opinion. Thus, the proportion of births and deaths in the upper and lower districts of the St Lawrence—those of Montreal and Quebec—compared with Great Britain, are, according to the best existing data, nearly as follows:—

		Births.	Deaths.
District of Montreal,	1844, .	1 in 20	1 in 51
District of Quebec,	... .	... 20	... 41
Whole of Lower Canada,	... .	... 21	... 53
Whole of England,	1848, .	... 33	... 45

Thus, while the births in Canada add yearly 5 per cent to the population, those of England add only 3 per cent; and while the deaths lessen the population in England by $2\frac{2}{5}$ per cent, those of Lower Canada lessen it only by $1\frac{9}{10}$. If we deduct these opposing numbers from each other, as in the following table, we find the yearly increase in the two countries to be—

	Lower Canada.	England.
Increase by births, . .	5 in 100	3 in 100
Decrease by deaths, . .	$1\frac{9}{10}$ in 100	$2\frac{2}{5}$ in 100
Yearly increase, . . .	$3\frac{1}{10}$ in 100	$\frac{3}{5}$ in 100

By natural increase, therefore, there are added to the French-Canadian population of Lower Canada four persons for every one that is added to the population of England. Not only are the people naturally prolific, therefore, but the country is pre-eminently healthy.

Another consideration will place this latter remark in a still stronger light. In England there died of the whole people yearly, in

1700,	one out of every		25
1801,	...	...	35
1811,	...	...	38
1848,	...	...	45

—that is, while England was undrained, and sanatory measures unattended to, and medical skill less extensively or generally available, the proportion of deaths was nearly double what it is now. Were it still as great as it was in 1700, and the births not more numerous, four out of every hundred would die yearly, while only three to every hundred are born. The population, that is, would decrease more rapidly than it is at present increasing. But Lower Canada, in its natural condition — undrained, innocent of sanatory associations or commissioners of sewers, and sparsely provided with medical men—is more healthy than England, with all the appliances which wealth, science, and civilisation have yet brought to bear upon her sanatory condition.

This is a very satisfactory result, not only as it explains why the French population of Lower Canada rapidly increases, but because of its bearing on the future prospects of the province, and the chances of health and longevity to those who may select it as their future home.

Among other agricultural observations I had made to-day and yesterday, I may mention the numerous small wind-machines, for grinding their corn, which I had seen attached to the barns of the habitants. Farms of 80 and 160 arpents possessed them; but I had not the satisfaction of seeing any of them in operation. Many of them looked old, and of somewhat primitive construction, so that the introduction of these mills must be of comparatively ancient date. The large mill of a seigneur comes to appear as a real advantage to a neighbourhood—even though the *moultre* be heavy—when we see how many other machines it supersedes, and how much labour and anxiety it saves.

Kamouraska, at the mouth of a river of the same name, and a county town, is rendered lively in summer by sea-bathing visitors. It is considered one of the healthiest places in Lower Canada. Opposite to it are a few rocky islands, of little value, between which and the mainland the bed of the river is nearly dry at low water. From the shore to the mountains a flat of recent deposits extends, studded at intervals with knolls of granite and gneissoid rocks, almost destitute of soil, and covered, like the rocks of Scandinavia, with dwarf and stunted pines. If any doubt remained as to this flat having been anciently submerged beneath the level of the St Lawrence, it would be removed by the appearance of the islands, and the almost dry inter-channels which lie opposite to the modern town. Lift up the present bed of the river some fifty feet more, and a new stripe of similar land, studded with similar rocky eminences, would be added to the existing land.

I had travelled through rain above and mud below for several hours, when, about 5 P.M., I reached the river Du Loup, twelve miles beyond Kamouraska, having accomplished only eighteen leagues during the day. I found the inn very comfortable, however; and as I was informed that I must proceed nine leagues farther before tolerable quarters were to be obtained, I made up my mind to stay here, and recruit for a long day on the morrow. I had soon a nice fire in a clean room, and a well-arranged dinner-table before me; and I felt very grateful to the good people of Quebec for the summer patronage they bestow upon the place during the season for bathing, since to a careful provision for their wants I was chiefly indebted for the well-served table and delightful night's quarters which I enjoyed at this hotel.

Oct. 3.—Du Loup, at the mouth of the river of that name, is a village of about a thousand inhabitants: it contains many good houses, a good hotel, and a larger proportion of British settlers than any other town in Lower Canada, with the exception of Quebec. It owes its actual prosperity, and the promise of future increase and importance, chiefly to its position. At the mouth of a small river, it has a harbour into which vessels can enter, and by means of which foreign traffic can be maintained. It is also at the end of the Grand Portage, as it is called—the highroad between Lower Canada and New Brunswick. The road to Canada, of which I have already spoken—as commencing at Little Falls, or Edmonston, on the upper St John River, in Madawaska, New Brunswick—leads along the Lake Temiscouata in a northerly direction, and reaches the St Lawrence at Du Loup, a distance from Little Falls of seventy-six miles. As the port not only of a large district of Lower Canada, but also of the upper waters of the St John River, and the more northerly portions of Western New Brunswick, this little place, as the population increases, and facilities of communication extend, must constantly increase in importance. At present also it is, in the summer season, the resort of sea-bathing parties from Quebec, whom the change of air, the romantic scenery, the beautiful views of the river St Lawrence, here upwards of twenty miles wide, studded with islands, and skirted on its opposite shore by lofty mountains, and the facilities of a steamboat which, in the hot months, plies from Quebec, allure to this, in British ideas, comparatively distant spot.

About half-past seven A. M. I left Du Loup, keeping still along the south shore of the St Lawrence. The morning was gloomy, and the air oppressed with fog, but fortunately the rain kept off for the greater part of the day. I found the habitants, with whom I engaged for horses and waggons from this point, each anxious to carry me farther than his horses were able to go. Though badly fed, indeed, it is astonishing how much work these active little horses can do. They rarely receive oats, unless they are to be put to unusually hard work; and, by way of preparation for the long journey of this morning, my first driver had overfed his horse, and actually unfitted him for a fair day's work. Notwithstanding bad and heavy roads, however, he persisted in whipping on till we had accomplished twelve leagues, when, after frequent remonstrances on my part, he was himself compelled to confess that his horse could do no more, and I unwillingly to send him back. The next horse carried me ten leagues, and, about eight P. M., brought me to the town of Rimouski, where I with difficulty found quarters for the night.

The road from Du Loup lay through a thickly-peopled country as far as St Simon, and a few miles beyond; it then entered upon a new clearing, called the Portage, along which, nearly the whole way to Rimouski, scarcely a single house existed twenty years ago.

Cacona, a rocky peninsula, about ten miles below Du Loup, which we passed on our left, rivals the latter as a sea-bathing station for the people of Quebec. It has the same views of the St Lawrence, and a salter water, but is not so romantically situated as Du Loup. Opposite to it is Green Island, on which stands one of the few lighthouses with which the St Lawrence has hitherto been provided. It is lighted up from the middle of April till the middle of December.

I have said that, for a considerable part of my day's journey, until I reached what is called the Portage, I passed through a thickly-peopled country; and so it really seems to the traveller who drives merely along the high-road, and judges, as one is inclined to do, only from what he sees. To pass house succeeding house, at distances of one or two fields only, for a whole morning, creates the impression of a dense population of small farmers; and although the woods appear to close in behind each small farm at no great distance, yet knowing that each farmer keeps many acres in wood for his winter's fuel, we naturally conclude that beyond these reserved woods other farms are cleared and cultivated in numberless Concessions behind. But this is the case as yet only to a small extent; and the vast area of the still untouched Lower Canadian wilderness, as well as the comparatively small fraction of the surface which has yet been subdued by the hand of man, may be judged of from the following table. The four counties of Bellechasse, l'Islet, Kamouraska, and Rimouski are those through which I passed, on my way from Quebec to Mitis, along the south shore of the St Lawrence. The area of these four counties is represented in square miles and in acres in the first and second columns; and the extent of each actually surveyed as yet, and laid out in square miles, is indicated by the numbers set down in the third column.

	Area in		Laid out in
	Square Miles.	Acres.	Square Miles.
Bellechasse, .	1083	693,020	726
L'Islet, . .	1220	730,800	560
Kamouraska, .	1090	6,976,000	568
Rimouski, .	8200	5,248,000	2240

In the first of these counties, that opposite the Falls

of Montmorenci, about three-fourths of the whole surface has been surveyed, and probably, for the most part, granted. In l'Islet and Kamouraska, about one-half has been surveyed; but of that of Rimouski, only one-fourth. Of that which has been surveyed, probably a large portion is still ungranted; and of the granted, at least two-thirds is still uncleared. But in Rimouski alone, there are still four millions of acres of unexplored, and, except to the lumberers, almost unknown wilderness. In this county, in 1848, about 8,000 acres of land were sold, one-third of it to new settlers; and the value of cleared land, within the last ten years, has increased 15 per cent. Even this remote county is advancing, therefore; but, at this rate of increase, how very many years must pass before all the available land can be subjected to the plough?

But the rapidity of settlement even where, as in these counties, it chiefly depends on the natural increase of the population, advances in a geometrical progression. Land and subsistence are easily obtained on the seignories, and the climate is healthy; the habitants marry young, and have large families. The clearings through which I passed to-day were all made by French Canadians. My driver was one of fourteen children—was himself the father of fourteen, and assured me that from eight to sixteen was the usual number of the farmers' families. He even named one or two women who had brought their husbands five-and-twenty, and threatened "le vingt-sixième pour le prêtre." I expressed my surprise at these large families. "Oui, Monsieur," said he, "vous avez raison. Nous sommes terribles pour les enfants!"

The eldest son usually succeeds to the father's farm; the younger sons are *placed* upon new farms, which the father helps to buy or clear. The rate at which the Government land is sold is half-a-dollar an acre—thirty or forty dollars for a farm one acre broad and thirty or forty deep—four years being allowed to make the payment. The seigneurs in this district concede for a yearly reserved rent of half-a-dollar for the lot, and one-twelfth of the value at each change of ownership.

But it is not all parents who are thus provident for their children, or in all localities that land can be obtained so near the paternal home that the children shall not object to remove it. According to the French law, the father's property is shared out among his children when he dies; and thus, in many cases, instead of leaving the home-farm wholly to the eldest son, the family of sons parcel it out among themselves. Four sons will divide a farm of two arpents in front, and thirty or forty backwards, into four long stripes of half an arpent broad in front, and thirty or forty in length. Thus the evils necessarily attendant upon the original shape of the farms become manifold increased; the *morcellement* proceeds, unhappily, in some localities, as it has already done in so many districts of France and Belgium; and the poverty of the people advances in proportion. It is the exact counterpart of the subdivision into long stripes which has led to such evil results among the sub-tenantry on many estates in Ireland—a similar Celtic population.

The effect which such a subdivision, followed by the building of houses along the road-side upon each lot, must have upon the apparent populousness of the country, in the eye of the traveller, cannot fail to strike the reader.

The system of clearing adopted in Lower Canada seems to be nearly the same as I had already seen in New Brunswick; and the log-cabins upon the new clearings are quite as poor, and as poorly provided, as those of the Irish immigrants into that province.

In this day's ride, however, I was struck with the

apparently very unpromising places in which some of the new settlers had built their log-houses, and were beginning to clear the land. In rocky spots, and along damp boggy hollows, where little available land appeared, the love of a neighbouring home, I suppose, of society, and of being near their relations, had caused many to select their farms and fix their future dwellings. The solitude of the remote clearing, where the land is better—into which even the Irishman with his family, supported and sustained by his passion for the "bit of land," will boldly venture—is unsuited to the less energetic and self-dependent habits of the French Canadian and his superstitiously-religious feelings.

The monuments of their industry, however—as in some of the rocky parts of Ireland—are every now and then conspicuous in these unpromising places. The piles of stones, of all sizes, I saw collected here and there, as I drove through the clearings, spoke strongly not only for their industry but for their perseverance. The fields, when thus cleared, often afford good land; but the expense of labour in clearing these unpromising spots must be enormous, and is quite beyond the ability of immigrant settlers from a foreign country.

I was struck to-day with a general resemblance in the outward appearance of this people to our poorer Irish. The broken panes in the windows of the houses were stuffed with old hats, and the clothes on the backs of the peasantry were often in tatters. On the other hand, the stylish French character of the improved houses, whitened over with quicklime, and the evident desire everywhere to build these better houses, showed an apparent aversion to live in misery and filth, for which the humbler Irish are by no means distinguished. Even here, however, the likeness may be thought to hold. These externally fine houses are anything but clean and comfortable—according to our notions—within; and the love of this kind of display too often leads the poorer farmer to spend upon a house what he must raise by a mortgage upon his farm, and is frequently the cause of his losing both house and farm to a pressing mortgagee, and being compelled to begin the world in a log-house anew.

Among the matters of social economy, indeed, which have struck me most, not only in the British provinces, but in New England and in the eastern States of the Union, is the very wide extent to which, according to the information I received, the property of these small farm proprietors is mortgaged. The surest way of obtaining possession of a coveted farm is to advance money on it by mortgage, as, in a very large majority of cases, the unhappy borrower is unable to pay off his debt. It may be the consequence of the failures of crops of late years, or of that general imprudence of the farmers in the lumbering districts to which I alluded in my former chapters on New Brunswick; but that these records of debt and difficulty among the farming community of North America generally are at the present time very numerous, there is no reason to doubt.

This Rimouski district is a lumbering country still to some extent, and, in consequence, labour is dearer than in the other counties. From £18 to £25 currency, with board and lodging, are the wages of a farm-servant; while carpenters and other handicraftsmen receive a dollar a-day. That wages should be so much higher here than they are a couple of days' journey up the river, shows either how unwilling the natives are to leave their own neighbourhood, or how little is the intercourse which exists between the inhabitants of the different counties.

The whole drive down the St Lawrence—in fact, the course of the river itself—is along the strike of the older Silurian beds. The country which the road traverses consists of flats along the shores of the river, or of longitudinal valleys between opposite hilly ridges, more or less rocky, from between which the edges of softer beds have probably been scooped out before the latest upliftings of the whole land took place. Inland the country consists, for a great distance, of alternate ridges and valleys—a prolongation of the mountains of Vermont and New Hampshire.

In occasionally passing from one of these valleys to another, we ascended and travelled for some distance along the upland. Where the slates happened to be soft and crumbly, the soil of these uplands was almost always good; while it was rocky, stony, or light, and of little depth, where the slates were harder, or of a more metamorphic character. In the bottoms of the valleys, as in this climate we should expect, bogs were frequent—savannahs, as they are here called—of a deep black earth, formed in the same way as our peat-bogs, but different in the physical qualities of the material of which they consist. The fibrous tenacity of our peat is wholly wanting; a spungy but crumbling black vegetable mould is the almost universal material of the North American bogs—unfit for the manufacture of peat, but of great use to the farmer, and highly valued for the preparation of fertilising composts. The absence of the heaths and mosses, of which our bogs are formed, and the roots of which long remain undecayed in the brown, and even in the black peat, together with the greater extremes of summer and winter temperature to which the decaying matter is exposed in North America, are probably the main causes to which the difference in the physical qualities of the bog-stuff in Europe and America are to be ascribed. Travellers in New England, who are interested in agriculture, will be surprised to hear this black bog-earth universally spoken of in conversation, and in agricultural books, under the name of *muck.*

It is a mark of the condition of agriculture in these countries, that the old word muck, (never mentioned now to ears polite,) which means the same thing as manure, should have obtained a specific application to this black earth. It shows not only how much in general has been neglected among them, but to what kind of muck of nature's making, when they at length found it necessary to apply something to their land, they were content to intrust the safety and productiveness of their crops.

It was already becoming dark when I reached the town of Rimouski, and I was driven up in succession to several French houses where strangers were said to be received, but was dismissed unceremoniously from one after another. At length I pulled up at a shop kept by an Irishman, where strangers were accustomed to lodge; but here, too, I was told that no entertainment could be had. Through the kindness of my friend, Professor Horan, of the seminary at Quebec, I had been provided with a letter of general recommendation to the Roman Catholic clergy on my route, which I meant to use only in case of an emergency arising; and I was about to order my *cocher* to drive to the *presbytère*, which was large, comfortable-looking, and near at hand, when a female came to the door of the house. The sound of an English tongue in her ears soon smoothed all difficulties, and I was speedily admitted, and made as welcome as her means would admit. But her husband was a sad screw, and provoked me much by the detention he caused on the following morning, in failing to procure for me, according to promise, the means of advancing on my journey.

The reason of my being bandied about from house to house proved to be, that my French companion took me from one French house to another, where they refused to receive me because I was English; and then, when I got to the Irishman's house, I was refused again, because, till I began to speak for myself, I was believed to be French.

Oct. 5.—After an unpleasant delay in procuring a conveyance, I was again on my way from the little town of Rimouski towards Mitis. A tolerably good road—the latter part of it close to the river—a good horse and an intelligent driver, made the morning's drive agreeable, and I reached Mitis about one o'clock. By the way were many French settlements scattered at intervals, and now and then a superior-looking house. My driver, himself a thriving farmer, informed me that here, as elsewhere, his countrymen often built houses beyond their means. Passing one of more than ordinary pretension, which had cost £500, he remarked that the owner had contracted much debt in building it, and that, when his farm came to be sold, it would not bring more than the sum which the new dwelling-house alone had cost.

Around Mitis, which is not even a village, there is much good hardwood—maple-land. About thirty years ago the first house was built in this neighbourhood, and now there is a large settlement of prosperous farmers, the greater part of them Scotch, who settled here a number of years ago. Few settlers from Scotland have come to this place during the last few years; but the habitants are flocking to it from the higher parts of the St Lawrence, partly because the land is good and crops excellent, and partly because some of it at least can be obtained in freehold. The greater part of the land around Mitis is held from the seigneur, but the Government also possesses land which is sold at from 1s. to 4s. an acre. From the conversations I had with the natives, they seemed to value highly the freedom from annual payments, and from the *droits de vent*, or as they expressed it, "being the seigneurs of their own land."

Mitis lies at the northern extremity of what is called the Kempt Road, which connects the waters of the

Restigouche with those of the St Lawrence. Along this road there are as yet scarcely any settlements, the population of the Lower St Lawrence being chiefly confined as yet to a stripe of a few miles broad along the river. The extra produce in grain, &c., is shipped to Quebec, from whence all necessary supplies are obtained in return. A merchant located at Mitis, and with whom I took up my quarters, serves as the medium of communication between the farmers of the district and the importers of Quebec. In autumn he gathers in his debts, in the form of produce, from his neighbours; and in return for these, obtains his winter's supply of tea, coffee, and clothing from the capital of the province.

These supplies, during the winter and spring, he again sells chiefly on credit, and waits for his payment till harvest comes. The system is worse for the farmer than the merchant, whose profits are large.

I found it necessary here to engage a horse and light waggon to take me all the way across the peninsula traversed by the Kempt Road, a distance of eighty miles, as neither horse nor conveyance were likely to be obtained by the way. I was glad to find that nothing was said as to the practicability of conveying myself and my portmanteau along this route, which my friends in New Brunswick had assured me I should find next to impossible. But difficulties always lessen when you look them fairly in the face; and I had afterwards occasion to find that, in regard to many other things having a relation to their own country, the New Brunswickers knew quite as little as I did myself.

I found it impossible, however, to arrange for proceeding further to-day, and was therefore obliged to postpone my departure from Mitis to an early hour to-morrow morning.

CHAPTER XIII.

Ideas generally entertained of American fertility and agricultural resources.—Reports of travellers.—Desire to obtain accurate information.—Condition of agriculture as an art in North America.—Contrast between Europe and America.—General effect of an exhausting culture upon the soil.—Effect sometimes produced very slowly.—Instance of old abbey-lands.—Claim of monasteries to the manure of their tenants' stables.—Effect of general exhaustion on the production of staple crops.—Its effect on the wheat-lands of North America.—Their retreat towards the west.—Liability of plants to disease on impoverished lands. — Remarkable change of cultivation in Lower Canada during the last twenty years.—Great diminution in the wheat and increase in the oat crop.—Loss and disaster which must have accompanied such a change.—Effect of this change on the corn-markets of the world.—Lower Canada become wheat-importing and oat-consuming. — Disastrous effect of the potato failure.—Similar changes threaten to follow similar modes of culture in other parts of North America.—The wheat-exporting capability will diminish—Manuring system of Scottish farmers who sell or carry off their crops as is done in America. — Possible continued and extensive supply of Indian corn.—Import-duty in the United States upon corn from Canada: should it be removed?—Would it on the whole be beneficial to Canada?—Zeal for improvement in Upper Canada.—Why do Rochester millers compete with the Canadian in Liverpool, in flour made from Canadian wheat?—Occasional low freights of the New York liners.—Use of Canadian wheat for mixing. — Alleged large mercantile profits expected in Canada,—Profits derived from dealings in land.—Profit of a direct trade in flour between Montreal and Liverpool.—Growth of flax in Canada, and export of linseed.—Instance of the close relation of discoveries in science to the profits of agriculture, and the agricultural capabilities of a country.—Comparative freights by the St Lawrence to Liverpool, and by the Erie Canal through New York.—St Lawrence the natural outlet of the lake-bordering countries. — Great expertness with which the Erie

Canal is worked.—Overflow of traffic upon that canal.—Great exertions of Canada in the construction of canals.—The Welland Canal. —Canals along the rapids of the St Lawrence; their extent and cost. —Energy of Canada compared with that of New York.—St Lawrence route now takes less time than that by the Erie Canal.—Money cost of transport is also less.—Ohio wheat will take the St Lawrence instead of the Mississippi route. — Importance of this route to the political independence of the free North-western States.—Difficulties in the navigation of the Lower St Lawrence.—Structure of its bed and channels.—Risks, and high insurance.—Need of lighthouses and depots of stores at the mouth of the river.—Traffic by the Richelieu Canal and Lake Champlain.—Rising importance of this.—Projected new ship-canal.—Future prospects of the St Lawrence navigation.

Before I leave the shores of the St Lawrence, there are a few points in connection with the agriculture and national economy of Canada which have been the subjects of observation and of interesting consideration to myself, and which most of those readers who have accompanied me thus far in my book will not regret to have their attention drawn to. Until I personally visited North America, my own notions as to the agricultural condition, capabilities, and resources of the several new provinces and States, were, I now find, notwithstanding all I had heard and read, of the crudest, most general, and indefinite character. The exaggerations of interested natives and settlers, and the repetition of such exaggerations by travellers who knew nothing of agriculture themselves, and, like myself some dozen years ago, could scarcely distinguish bad land from good—these were all the information our journals and yearly literature afforded us. That wheat and Indian corn poured in upon us at times from those regions, we knew; that some portions of the country were rich and fertile, we could not doubt; and the general conclusion in the public mind was, that these new countries were generally fertile—that inferior land was the exception—that large crops were everywhere reaped—that the fertility of the whole region was inexhaustible — that the supply of wheat it could send us was without bounds—and that if those who tilled the land and raised the corn in these countries were not so skilful as the average of our own farmers, this was only another evidence that nature there was kinder to the tiller of the soil than she is in our own country, and did not demand at his hands either the same amount of knowledge, or the same unwearied application of ceaseless toil.

One of my objects in visiting North America was to remove the mistiness of my own ideas as to the agricultural character and condition of its several great regions, to test the seeming exaggerations in which, as if by some natural law, the natives and residents of this northern part of the New World are inclined to indulge. I was desirous, also, of obtaining a clear idea of the relation which American practice bears to English practice; the prospects and success of individual American to those of individual English and Scotch farmers; American past and future surplus wheat to the state and demands of the English market; the life of the settler in these new countries to the life he would have led had he remained at home. On a few of these points I have arrived at clear and definite notions—not hastily, I believe—though some of them may still be incorrect. It is some remarks upon these I wish briefly to put down in this place.

And first, as to the condition of agriculture as an art of life, it cannot be denied that, in this region, *as a whole*, it is in a very primitive condition. Before the first Puritan emigrants landed at Plymouth, the Indians planted and hoed and reaped their corn much as the white settlers do now, and, like them, deserted old land for new when the crops began to fail. Many operations, it is true, are now performed upon existing farms which were unknown to the Indian races; but a similar absence of skill and forethought is generally observable in reference both to the mode of performing them and to their after effect upon the land. I speak, it will be borne in mind, in these remarks, of the newly settled parts of North America; and the more newly settled the more closely will they apply. I would not be understood to calumniate those longer cultivated districts in which—the first stage of their agricultural history being past—new life and energy are now being brought to bear upon the culture of the land, and by which the errors of past ignorance and want of skill may by-and-by be repaired, or of those happier new districts which men of knowledge and capital are redeeming from a state of nature, and at once submitting to a rational system of culture, capable of being carried on for an indefinite period without injury to the land.

Were the population as fond of their homes, and as stationary in numbers, as in the central regions of northern Europe—as quiescent in character, their labour as small in money-value, and everywhere as abundant, and their institutions as repressive of exuberant energy—this primitive condition of the agricultural practice would be both less felt among themselves, and be a matter of less observation to foreign countries. As in Poland and Russia, land which had become unproductive would be abandoned to nature, and new land broken up year by year. Thus supplies of food comparatively uniform would be yearly reaped, sufficient not only to meet the wants of the native population, but to pour a large annual surplus into foreign markets, wherever a demand might happen to exist.

But all these things are different in North America. The population increases rapidly, not only by natural growth, but by the crowding in of immigrants from the various countries of Europe; and thus the mouths are greatly multiplied which are to be filled by the native produce of the land. Labour is comparatively dear, so that new land cannot be cleared and brought into cultivation as fast as the old is worn out, were new land even everywhere available, which in the more settled parts of the country, already divided into small farms, it usually is not. Besides, the majority of those who boast of Anglo-Saxon blood are generally energetic, and their institutions incline them to push everything forward as in a race—their wide continent holding out to them many dazzling hopes and sources of gain. They labour, therefore, those who till the soil, to make as much and take as much out of the land as they can, and in the least possible time; probably without either thinking or wishing that their actual residence is to be the future home either of themselves or of their children, but rather that interest or expediency may by-and-by carry them all to happier homes in the farther west.

The result or effect, therefore, of this condition of the rural art, and of the agricultural population, upon the state of the soil, is to bring it by degrees into a state of more or less complete exhaustion. Whatever be its quality or natural fertility, this is the final and inevitable result. In land which is very rich, the effect is produced more slowly—so slow, that those who hold land which for fifty or a hundred years has yielded crops of corn without the addition of manure, will scarcely believe in the possibility of its ceasing at last to give its wonted returns. But old experience and modern science alike demonstrate that the richest soils, by constant cropping, without the addition of manuring substances to replace what the crops carry off, must ultimately arrive at a state of comparative barrenness.

It is not to be wondered at that men should be faithless upon this point, when it is considered how grateful the soil is for kind treatment, and how very long, in some cases, it is before it begins to resent a contrary course of procedure. The lifetime of one man may be spent in gradually improving and enriching a field by skilful management, and the whole lives of two successors may be employed in impoverishing it again without reducing it to the low condition from which it had originally been raised.

We have in England many remarkable instances of this latter fact, which, in the neighbourhood in which they occur, are often considered inexplicable. Near the ruins of old abbeys and monasteries, for example, which have been long demolished, a few fields are often observed which exceed in fertility all the adjoining land, and, though treated no better than the rest, have ever within the memory of man given better crops. Facts of this kind are to be explained by a reference to the customs of the time when the ecclesiastical buildings still flourished. It was the habit of the officials of the church or monastery to collect, as a due, all the manure made by the cattle and sheep of the cottars or tenants who held under it, and to apply it to the home farm, which was cultivated for the immediate use of the church. The holders of land, by way of rent, for every four acres they held, were in some instances bound to plough one acre for the church; and to this, and to the grass land kept in their own hands, the servants of the church applied the manure which they took from the tenantry.* It cannot,

* Illustrations are to be found in the following extracts from the *Chronicles of Jocelyn of Brakelond*, pp. 29 and 30 :—

"And it was done accordingly, and confirmed by our charter, that there be given to them another quittance from a certain customary payment, which is called *sorpeni,** for four shillings, payable at the same term. The cellarer was also used to receive one penny, by the year, for every cow belonging to the men of the town for their dung, (unless, perchance, they happened to be the cows of the chaplains or the servants at the court lodge ;) and these cows he was used to impound, and occupied himself much in such matters. . . . . . .

"The cellarer was to have the ploughings and the other services—to

* *Sorpeni.* This word is the same as scharpenny or scharnpenny—that is, dung-penny, from *schearn*, dung. By this it seems the tenants were bound, as being originally bondmen, to pen up their cattle at night in the pound or yard of their lord, for the benefit of their dung; and if they did not do so, they paid this dung-penny as a compensation.

I think, be doubted, that the differences still observed in the fertility of lands adjoining the ruins of monastic houses, are in part to be ascribed to the long continuance of a practice by which the natural fertilising substances of a whole neighbourhood were lavished on a comparatively restricted space. That it is very long since this marked distinction was made in favour of the abbey home-farm, is the point I wish to bring out. It shows that land is very grateful, and that when once enriched, either by nature or by the hand of man, it may long resist exhaustion, and maintain a decided superiority over that which adjoins it. But its doing so is no proof that, though later, it will not also be finally exhausted, if submitted to inconsiderate and selfish modes of culture.

The first practical or economical consequence of this exhaustion of the land is, that it gradually ceases to produce a remunerative return of those crops which have been specially cultivated upon, and have been the immediate means of exhausting it. In North America, generally, this crop has been wheat—as this has always been the kind of grain for which the most ready market could be obtained, or which could be most certainly exchanged for the West India produce and the manufactured articles which the settler required. As the exhausting culture

wit, the ploughing of one rood for each acre, without meals, (which custom is still observed,) and was to have the folds where all the men of the town, except the steward, who has his own fold, are bound to put their sheep, (which custom, also, is still observed.) . . . . . .

"Also, the cellarer was used freely to take all the dunghills in every street, before the doors of those who were holding *overland;* for to them only was it allowable to collect dung, and to keep it. This custom was not enforced in the time of the Abbot Hugh, up to the period when Dennis and Roger of Hingham became cellarers, who, being desirous of reviving the ancient custom, took the cars of the burgesses, laden with dung, and made them unload; but a multitude of the burgesses resisting, and being too strong for them, every one in his own tenement now collects his dung in a heap, and the poor sell theirs when and to whom they choose."

proceeded, therefore the quantity of wheat raised beyond the demands of the state or colony—that is, the surplus for exportation—gradually decreased.

I need not enter into details upon this point; the grand consequence is such as I have described, and the general proof of it is, that the wheat-exporting regions of North America have, as I have already stated in my remarks upon western New York, been gradually shifting their locality, and retiring inland and towards the west. The flats of the Lower St Lawrence were the granary of America in the times of French dominion; western New York succeeded these; next came Canada West; and now the chief surplus exists, and the main supplies for the markets of Europe are drawn from the newer regions beyond the lakes. These in their turn, as the first virgin freshness passes away, will cease to be productive of abundant wheat, and eastern America must then look for its supplies of this grain either to a better culture of its own exhausted soil, or to regions still nearer the setting sun.

This natural consequence of an exhausting system of culture has been aided and hastened by other causes, the study of which is full of interest and instruction. I may advert to one of these.

When a soil becomes unfavourable to the growth of a plant, the plant, if made to grow upon it, comes up weak, and is liable to disease and to the attacks of insects and parasitic plants. Whether as a natural consequence of this kind, arising naturally from the exhaustion of the soil and the weakening of the wheat-plant, or as the effect of some other cause not understood, it is an important fact that the attacks of the wheat-midge have, in Lower Canada, been lending their aid for many years to diminish the wheat-crop in quantity, and to render it less certain. A gradual revolution, therefore, has been taking place, not only in the husbandry, but in the food of the people also, and in the kind as well as quantity of the surplus produce they have been able to bring to market. I know, indeed, of no well-ascertained facts in the agricultural history of any country which are more striking in these respects than those which are presented by a comparison of the quantities—relative and absolute—of the different kinds of grain produced in Lower Canada, at successive periods, during the last twenty-five years.

The following table, published by the Canadian Board of Statistics in 1849, exhibits the amount of this produce in bushels in the years 1827, 1831, and 1844, respectively:—

	1827.	1831.	1844.
Wheat, . .	2,931,249	3,404,756	942,835
Barley, . .	363,117	394,795	1,195,456
Oats, . .	2,341,529	3,142,274	7,238,753
Rye, . .	217,543	234,529	333,446
Indian corn,	333,150	339,633	141,003
Buckwheat, .	121,397	106,050	374,809
Pease, . .	823,318	948,758	1,219,420
Potatoes, .	6,796,300	7,357,416	9,918,869

In this table we see that from 1827 to 1831, and probably somewhat later, a similar state of things existed, and that a gradual increase took place in the amount of all the crops raised; a natural consequence of the increasing population, and of the larger breadth of land every year subjected to the plough. The wheat-crop increased by 500,000 bushels, the oat crop by 800,000, and the potato crop by 500,000. In these quantities we see a slight tendency to an increase, in the proportion of oats grown, above that of wheat or potatoes; but in the other crops there is nothing to arrest especial attention.

In 1844, however, a very different state of things presents itself. During the interval of thirteen years (from 1831 to 1844,) the wheat-crop, instead of increasing 2,000,000 bushels, as it ought to have done, had diminished from 3,500,000 (its amount in 1831) to less than 1,000,000 bushels. The barley crop, on the other hand, had increased by 800,000 bushels, that of pease by 400,000, of potatoes by 2,500,000, and of oats by the enormous quantity of 4,000,000 bushels.

Whoever is acquainted with the practical operations of husbandry, will be able to conceive how many anxieties and losses, and repeated failures of usual crops, must have beset the unhappy farmer, before his course of cropping could be so changed as almost entirely to substitute oats for wheat in the fields he had set aside for grain. The wheat was clung to by the Canadians with the more pertinacity, because it was the crop which brought in the annual supplies of money and other foreign articles, and because it formed a considerable part of their usual food. The failure of this source of supply brought debts and mortgages, and transference of property; and to it is to be ascribed a considerable proportion of the mortgages which, as I have said, hang round the necks of the rural population, over so much of this part of north-eastern America.

In relation to the corn-markets of the world, this change converted Lower Canada, on the whole, from an exporter into an importer of wheat—as it no longer produced enough for its own consumption; and in reference to a large part of its own population, which was unable to buy wheat, turned them from the consumption of this grain to that of potatoes and oats, with a lesser quantity of pease and barley. This was before the failure of the potato crop; and in this state of things, when, by the previous failure of the wheat, the potato had become doubly precious, it will be understood how the potato disease must have produced a more intense amount of suffering among the Lower Canadians. The French population naturally dislike the oat for food, and consume very unsightly and distasteful varieties of soft bread, rather than live upon oatmeal. But late years have loosened their prejudices very much; and both in Canada and New Brunswick the oat is now becoming an article of extensive consumption, even among the inhabitants of French descent.

The case of Lower Canada illustrates, in an exaggerated degree, what I believe is the natural consequence —perhaps we may say the natural sequence of events— in countries where the agricultural practice, for a series of generations, is such as it has hitherto been in North America generally. The staple crops—the supposed staff and agricultural strength of the country—first fall off, and then change; and with this change the food of the masses, and the relations of the country as a whole with foreign markets, change also.

This has already been the case in the longer settled portions of the North American continent; and the same consummation is preparing for the more newly settled parts, *unless a change of system take place.* The new wheat-exporting—so called—granary districts and States will by-and-by gradually lessen in number and extent, and probably lose altogether the ability to export, unless when unusual harvests occur. And if the population of North America continue to advance at its present rapid rate—especially in the older States of the Union—if large mining and manufacturing populations spring up, the ability to export wheat to Europe will lessen still more rapidly. This diminution may be delayed for a time, by the rapid settling of new western States, which, from their virgin soils, will draw easy returns of grain; but every step westward adds to the cost of transporting produce to the Atlantic border, while it brings it nearer to that far western California, which, as some predict, will in a few years afford an ample market for all the corn and cattle which the Western States can send it.

In their relation to English markets, therefore, and the prospects and profits of the British farmer, my persuasion is that, year by year, our Transatlantic cousins will become less and less able—except in extraordinary seasons—to send large supplies of wheat to our island ports; and that, when the virgin freshness shall have been rubbed off their new lands, they will be unable, *with their present knowledge and methods*, to send wheat to the British market so cheap as the more skilful farmers of Great Britain and Ireland can do.

If any one less familiar with practical agriculture doubts that such must be the final effect of the exhausting system now followed on all the lands of North America, I need only inform him that the celebrated Lothian farmers, in the immediate neighbourhood of Edinburgh, who carry all their crops off the land—as the North American farmers now do—return, on an average, ten tons of well-rotted manure every year to every acre, while the American farmer returns nothing.* If the Edinburgh farmer finds this quantity necessary to keep his land in condition, that of the American farmer must go out of condition, and produce inferior crops in a time which will bear a relation to the original richness of the soil, and to the weight of crop it has been in the habit of producing. And when this exhaustion has come, a more costly system of generous husbandry must be introduced, if the crops are to be kept up; and in this more generous system, my belief is, that the British farmers will have the victory.

I have spoken, the reader will bear in mind, of wheat only. Make it an object to the Central States to send their maize to England, instead of converting it into pork for the packers of Cincinnati; and, as I have not examined those States, I do not know what limit should be placed to the quantity they could continue to send us for many

* The Edinburgh farmer sells all off—turnips, potatoes, straw, grain, and hay. But he manures his turnips with thirty, and his potatoes with forty loads of manure, in a rotation consisting of potatoes, wheat, turnips, barley, hay, and oats.

years to come. But Indian corn will never become the staple food of our people; nor is it desirable that it should. The importation of this grain will not directly interfere, therefore, with the home market for wheat.

The view I have taken of this question has a material bearing upon other topics which interest us at home, but which are at present peculiarly exciting to the people of Upper Canada. Take the import-duty of 20 per cent upon provincial produce, when brought into the United States, for an example. Wheat, like all other unmanufactured articles, pays this duty. In present circumstances, the impost on this article does not concern Nova Scotia, New Brunswick, or Lower Canada. They are wheat-importing countries, and are therefore indifferent about it. But Upper Canada produces fine wheat, and has still a large surplus, which, as a nearer and more convenient market, the Upper Canadian merchants are anxious to introduce into Rochester and Oswego free of duty. I have in a previous chapter expressed my opinion that a free commercial intercourse, between the opposite shores of the St Lawrence and of the great lakes, is desirable on general grounds; but, in order to obtain this free intercourse, it does not appear to me that Canada ought to be over-anxiously urgent with the officials at Washington, or rashly to offer concessions which may ultimately prove to be far beyond the value of the reciprocal advantage they now wish to obtain.

In regard to wheat, there are two or three points which are worthy of careful consideration. First, the power of exporting wheat resides at present only in the newer and richer parts of Canada West. Supposing that the result expected by the wheat-growers and corn-merchants of this part of the colony were to follow from this wished-for free importation of grain into the United States—namely, a considerable elevation in the price of wheat—would it be for the good of Canada, as a whole, that the price of flour should, in consequence, be raised to that large tract of country, on the Lower St Lawrence, which has ceased to raise enough for its own consumption?

Again, if the conclusions be well founded that the existing system of culture is an exhausting one, and that the wheat-exporting regions are, in consequence, retiring more and more towards the west, the same fate awaits Upper Canada which has already overtaken the equally fertile State of New York. The surplus of wheat beyond the wants of the home population—the ability to export, that is to say—will gradually lessen, and, except in extraordinary seasons, will finally cease. It is true that a larger proportion—about 80 per cent, it is said—of the population of the Canadas is engaged in agriculture than in the State of New York, and is therefore producing food; but against this is to be placed the fact, that so large a proportion of these agriculturists are already purchasers of flour. The probability, therefore, is really great, that Canada, *as a whole*, will fall off in the production of wheat, in comparison with the wants of its population, quite as rapidly as the State of New York has done.*

It is quite true that Upper Canada can boast of much zeal for agricultural improvement, and many enlightened and anxious promoters of rural advancement—as

* I have stated elsewhere, in the text, that I speak with reserve of Upper Canada, as I have not had the opportunity of visiting enough of it to form a satisfactory opinion. But the view given in the text is much confirmed by a passage in the address of the president of the Agricultural Association of this province, delivered in September 1850. "The farms," he says, "on the whole line in the old settled townships, from Montreal to Hamilton, and round the banks of the lakes, rivers, and bays, for a space of eight hundred or nine hundred miles, with few exceptions, are what is termed in Canada 'worn out,' and may be purchased at from £3 to £10 an acre." Suppose that better culture can restore this land, yet wheat can never be raised on it as cheaply as at first.

the State of New York also can—and I would not, in the slightest degree, undervalue the results to which their labours and example may lead, but would express my belief rather that much good to both countries will grow out of their exertions. At the same time, it is a thing to be calmly weighed, whether the anticipated pecuniary advantage of a portion of Canada only, and for a limited time, is worthy of any large sacrifice, on the part of the whole province, which sacrifice will be permanent and irrevocable.

Lastly, when speaking of the Rochester mills and millers, I have alluded to the apparent anomaly, that the owners of these mills can buy wheat at Toronto, pay the 20 per cent on importing it into Rochester, convert it into flour, send this flour to New York, from whence it makes its way to Liverpool, and, at this latter market, can still compete with the Canadian flour made from the same wheat.

Now, the large flour-mills of Montreal and of Upper Canada—as I was informed by an English fellow-traveller on the St Lawrence, who professed to be cunning in mills, and had visited both for the express purpose of comparing them with one another—are much superior to those of Rochester and Oswego; and, as another competent authority assured me, they can grind flour 15 per cent cheaper than the mills of these two places. Further, the average freight to Liverpool, down the St Lawrence, is not so great as the cost of transport from Rochester to Liverpool. If England, therefore, be the market to which nearly all the surplus flour of the United States which finds a sale in Europe is first sent, surely in that market the Canadian, coming directly from his own ports, should be able successfully to compete with the merchant from Rochester, who brings his flour to Liverpool by way of New York.

It may, indeed, be answered, that the constant sailing

of the regular *liners* from New York, often with little cargo, affords frequent opportunities of sending grain and flour to Liverpool at a cost exceeding very little the ordinary Stevedor's charges;* but this cannot materially affect the average freight of the flour, as a whole, which is annually shipped from that port to England. Or it may be said that the Rochester and Oswego millers make use of the fine Toronto wheat for the purpose of mixing only, and can therefore afford to give a higher price for it than the Canadian millers, who use unmixed Canadian wheat for the manufacture of their flour. But we can scarcely suppose that parcels bought for mixing would seriously affect the wheat market of Canada as a whole; or, if so, that the Canadian millers cannot mix and use up different wheats as profitably as those of Rochester.

Allowing, however, their full weight to these and similar considerations, I have been unable to satisfy myself that if the exporters of lumber—whose traffic has of late fallen off, and occasioned discontent—instead of countenancing or exciting political turmoil, would push the direct home trade in corn and flour, in place of that in timber, an equal commerce on the whole, and equally profitable to the colony, might be still maintained with the mother country.

One of the obstacles which, so far as I have obtained the means of judging, stands in the way of the energetic opening up of a direct wheat and flour traffic with Great Britain, is to be found in the large returns of profit which the merchants look for, and to which, I suppose, they must hitherto have been accustomed. What I have stated in a previous page, as having been told me by a Rochester miller, that the wheat-growers and millers of the United States are unable to com-

* For loading and unloading.

pete with Canada in the foreign markets, I mentioned to an extensive miller whom I met with in Upper Canada, on occasion of his complaining that the Rochester millers came over and bought the fine white wheat of Toronto at prices which he could not afford to pay. On this he remarked, after some hesitation, that "the Canadian would have to learn to be content with a profit of 5 per cent, when he had hitherto been accustomed to 50!" Large profits and long credits are always the characteristics of an infantile and circumscribed commerce in a new country like this. Facility and rapidity of communication, by shortening credits, and enabling the same capital to go farther, make large reductions in the rate of profit consistent with equal gains, on the whole, to the enterprising and intelligent merchant.

I have had other occasions to feel surprise at the large profits expected in Canada from the investment of capital. It having been mentioned to me that a friend of mine in Lower Canada had expended £15,000 in the purchase of a seignory, and that the return was £1500 a-year, or 10 per cent on the outlay, I—naturally enough, as we should think in this country—expressed my surprise at so large a return from investments in land. An influential member of the Canadian Legislature, who was present, on the other hand, considered it small, was surprised that similar investments in England did not yield so much, and stated that, were he to invest £15,000 in land in his own district of Upper Canada, he should expect from it a return of £3000 a-year, or 20 per cent. In what way, whether by farming it himself, or by letting it or selling it to others, he was to realise this return, I did not learn; but the possibility of doing so, except in some special case, I very much doubt.

But that some persons, at least, are satisfied of the

possibility of profitably conducting a direct traffic with England in Canadian flour was proved by the fact, that, while I was at Montreal, a vessel from the Ontario, containing seventeen thousand bushels of the best Toronto white wheat, lay opposite Goold's Mills—one of those of which I have spoken as being superior to the average of the Rochester and Oswego mills — to be ground, on merchants' account, for the Liverpool market. If properly cultivated, this trade may, I think, make both the growers and grinders of Canadian wheat very indifferent as to the 20 per cent duties of the States.

Among the articles of export from Lower Canada, linseed is one which used formerly to occupy a not unimportant place, though now, so far as I can learn, the export of this grain to England has almost ceased. The French Canadians used formerly to grow flax extensively for home consumption; and most of the Lower Canadian farmers still raise enough to employ and clothe their own families. The diminution in the growth and export of seed may be owing, in some degree, to the gradual substitution of cotton for linen in articles for domestic use; partly to the general exhaustion of the soil, of which I have spoken; and partly to the growing taste for finer cloth, which will necessitate the growth of a finer quality of flax. The first and third of these causes are probably the most influential. Now, it is known to all flax-growers that hitherto a fine fibre has been considered incompatible with a strong, rank, heavy crop of flax, or with the ripening of the seed. Hence the taste for fine flax would cause the sowing of much seed, that the plant might spring up thick—the selection of poor or exhausted land that it might not come up rank, or grow tall and strong—and the pulling of the plant before the seed was ripe. The more these practices for the improvement of the fibre were followed, the less would be the quantity of linseed brought to market.

It is one of those advances which the arts owe to scientific research, and deserves the consideration of those who affect to despise, or altogether deny, the use of science to agriculture, that the new method (Schenck's) of steeping flax in hot water promises to render all these precautions unnecessary, to extract as fine a fibre from the rank coarse ripe flax-plant, as from the slender unripe plant hitherto privileged alone to yield the finer thread. This method of steeping is certain and constant in its results, and is performed in as few days as the old method required of weeks.

The general introduction of this method of manufacturing the plant will simplify the farmer's treatment of the crop, will enable him to cultivate flax as he does any other plant he grows, to reap a profit from it in proportion to its total weight, and, as in other crops, to ripen his seed either for home use or for exportation. It may regenerate the flax-husbandry in Canada, and revive, without exhausting the land, the ancient trade in the seed as an article of export.*

I have said that the average freights from the Canadian ports, direct to Liverpool and other ports in Great Britain, cannot be greater than the cost of transmitting produce from the shores of Lake Ontario, through the port of New York. This direct freight ought in reality to be less; and in a few years it will almost certainly become so. This statement naturally leads me to make a few remarks on the navigation of the St Lawrence, its importance to Canadian interests, and the influence it is destined hereafter to exercise on the general revenues of the Canadas.

* Canadian seed ought to be as good as Riga flax seed, of which 5000 barrels have been imported into Newry, and 15,000 into Belfast, during the present season.

The natural outlet of the vast region of North America, which is drained by the great lakes and their tributary streams, is, as I have already remarked, by the river St Lawrence. This was early recognised; but the natural obstructions which existed in the channel of this river, have, with other obstacles, hitherto prevented it from being so easily and generally available as it is now likely to become.

In the first place, the rapids and falls of Niagara prevented the passage of vessels between the lakes Erie and Ontario. Thus, the Lower St Lawrence was inaccessible to the rapidly-settling western portions of Pennsylvania and Ohio. The idea of a canal from Lake Erie to the Hudson, through the low country of western New York, was therefore suggested, and was finally entertained by the New York State Legislature. The Erie Canal was the result; and up to nearly the present time this canal has formed the high-road between the upper lakes and the Atlantic, and has been a source of great wealth, and the cause of a very rapid prosperity, not only to the city, but to the whole State of New York. As the western country was cleared, and its population increased, the traffic along this canal augmented in a degree which the most sanguine had never contemplated, and extraordinary exertions have been made, from time to time, to facilitate the traffic, and to hasten the passage of the vessels with which it is crowded. The degree of expertness to which the working of this canal has been brought may be judged of from the fact, that, in the single month of October 1847, 6930 lockages were executed above Schenectady, which gives less than 6½ minutes for each lockage, Sundays included.

But every year causes new increase of traffic, and new delay in the transmission of produce and merchandise, and larger quantities are, in consequence, detained over winter, when the frost has put a stop to the navigation of the canal. The great traffic, and already immense population of the Western States, and the mineral and other produce of the upper lakes, now demand other ways of access to the Atlantic and to Europe.

Meanwhile the Canadian authorities, those of Upper Canada especially, have not been idle. Indeed, I believe they have done more to promote internal water-communication than any State of the Union—I may safely say, than any country in Europe—considering the infancy of their country, the extent to which its material resources have been developed, and the actual amount of its revenue and population.

First, the Welland Canal has been constructed, by which a direct communication for large vessels is established between the lakes Erie and Ontario. Thus the borders of the upper lakes were connected by a single freightage with the ports of western New York and with those of Upper Canada, along the borders of the Ontario, and down the St Lawrence, as far as Prescott on the Canadian, and Ogdensburg on the New York side, below which places the first rapids on that river occur. Upon this great work about £1,400,000 currency have been expended ; and, though still incomplete, it is already yielding a revenue of £30,000 a-year.

Next, the numerous rapids on the river, between Prescott and Montreal, have been flanked by canals, shorter or longer according to circumstances, by which the transit for large and loaded vessels, either upwards or downwards, has been rendered easy and secure.

Of these canals, the most important are:—

	Miles.	Locks.	Cost.
Williamsbury, four short canals, .	?	6	£245,000
Cornwall, .	$11\frac{1}{2}$	7	75,000
Beauharnais, .	24	9	310,000
Lachine, .	9	...	350,000

Below Montreal, the works in Lake St Peter cost £75,000, and the harbour of Montreal itself £131,000. I do not include, of course, as executed by the province the Rideau Canal, between the foot of Lake Erie and the upper waters of the Ottawa, as this was executed by the Home Government. It is a hundred and thirty miles in length, has forty-seven locks, and cost £800,000. Though intended chiefly as a military work, it will prove of immense benefit to the future development of the natural resources of the more northerly parts of Upper Canada in the great basin of the Ottawa.

Altogether, on the execution of canals and river-improvements necessary to the direct navigation of the St Lawrence from the upper lakes to the Atlantic, upwards of £3,000,000 currency, or twelve millions of dollars, have been expended by the Legislatures of Upper and Lower Canada. This sum is not only large absolutely or in itself, but it is especially so, compared with the amount of revenue hitherto at the disposal of the provincial Legislature of the Canadas. When we consider, also, that the whole canal debt of the State of New York is under seventeen millions of dollars, while the Canadas have burdened themselves with a debt of twelve millions, we shall be willing to allow that the amount of energy displayed by the people north of Lake Ontario and of the Thousand Isles is not less than has been manifested even in the State of New York, nor their faith less in the future growth and greatness of their rising country.

The result of all these improvements has been, that more and more of the direct European traffic with the great lakes has been making its way every year down the St Lawrence, instead of by the Erie Canal—even while that canal has been still able to overtake the whole of the traffic. But now that it has become evident that this canal, however it may be enlarged, and however energetically managed, will soon be wholly inadequate to the demands of the western trade, the value of the St Lawrence, and the certainty of its being very extensively used, becomes every day more clear.

Other considerations also are likely to hasten this result.

For laden vessels coming down Lake Erie with cargoes for Europe, the two points of destination are either Buffalo, at the mouth of the Erie Canal, on the New York side—or Port Maitland, at the mouth of the Welland Canal, on the Canadian side. If the vessel make for Buffalo, its cargo must be transhipped, sent 364 miles by canal, and down the Hudson to New York, and be again transhipped at least once before it can be despatched to Europe. If it enter Port Maitland, it passes the canals without breaking bulk, and descends to Quebec in four days. Thence the same vessel may proceed direct to Europe, or the cargo may be transhipped, and with a fair wind may pass the banks of Newfoundland before it could reach New York by the way of the Erie Canal. Thus, independent of possible detention on this canal, it appears that time is saved by the St Lawrence route; and every merchant knows the value of this element in commercial affairs.

Again, the cost of transport from Albany to Buffalo is $7\frac{4}{5}$ dollars per ton, while from Montreal to Port Maitland, ascending the river, it is only three dollars a ton—and the difference is greater in descending the river; so that the St Lawrence is also a cheaper route than that by Lake Erie. A fellow-passenger of mine across the Atlantic informed me that, in bringing railroad iron from Liverpool to Cleveland in Ohio, (on Lake Erie,) he found that, independent of speed, the route of the St Lawrence was 10s. a ton cheaper than any other he could take.

As less cost both in time and money, therefore, attend this route, it appears certain that the traffic will be increased upon it, not only by the surplus which the Erie Canal cannot convey, but by the diversion from that canal of a portion—perhaps a large portion—of the traffic it has hitherto monopolised. In other words, whatever the amount of traffic may be, the river St Lawrence will henceforth be able to compete successfully with the Erie Canal.

But this greater cheapness of transport, and the facility of establishing direct communication, and without transhipment, between Cleveland, Detroit, Chicago, &c., and Liverpool, will draw also into this eastern channel a large traffic which never sought Lake Erie, but made its long and tedious way down the Ohio and the Mississippi. The wheat and other produce of the valley of the Ohio, which was intended for the European markets, has hitherto, for the most part, descended those rivers, and, after a voyage of some thousands of miles, has reached New Orleans, whence it was reshipped to its European destination. But this long water-carriage, in the hot and humid climate of the regions through which these rivers flow, is found to affect the quality of the wheat; so that it rarely reaches Europe in so good a condition, or realises so high a price, as similar wheat does which has been conveyed through the Eastern States to the shores of the Atlantic.

It is easy now, however, to transport by railway to the harbour of Cleveland (on Lake Erie) the produce of the Ohio Valley; and as soon as it is generally known that the passage to Europe from that port, by way of the St Lawrence, is not only much shorter in time, but is also cheaper in money-cost, and brings the grain to market in better condition, it is clear that a portion, at least, of the European commerce with the Ohio Valley will be diverted into this channel.

I only allude to another circumstance and possible event which may dispose the free North-western States to cultivate this eastern route of the St Lawrence. At present, their main communication with the European markets is by the Mississippi, the mouth and keys of which are in the hands of the slave-owners of the south. Should a crisis between the free and slave States arise, this channel of intercourse might be shut up, and the prosperity of the States on the upper river and its tributaries thus greatly arrested. But if, in the meanwhile, this channel by the St Lawrence be cultivated, any contingency of that kind would fall less heavily upon the States affected by it, and the dread of its occurrence would seriously influence neither their policy nor their prospects.

To this view of the St Lawrence navigation, as bearing upon their own future condition and independence, both political and commercial, the free North-western States cannot be indifferent, and they will, therefore, be anxious, I think, both to promote it generally, and to secure its advantages to themselves as cheaply, and at as early a period, as possible.

Altogether, therefore, the prospects of the Canadian canals, and the general St Lawrence navigation, are exceedingly encouraging; and though, as I have said, I would not advocate, in these times, the creation of exclusive advantages on either side of the river and lakes, yet those which Canada now possesses ought not, with these bright prospects, to be lightly resigned to the United States, or without receiving a satisfactory equivalent.

But the Lower St Lawrence presents difficulties to navigation not yet alluded to, and which demand, at the hands of the Canadian Legislature, still further exertions, with a view to its improvement, and to the more effectually securing of the advantages which the commerce of the river is destined to confer upon the province.

The geological structure of Lower Canada, to which I have already in general terms alluded, affects the navigable character of the St Lawrence below Quebec. I have described the strike of the beds on the south side of the river as running nearly parallel with the course of the river from Quebec to the district of Gaspé, and the inland country as consisting of parallel ridges of more or less hard rock, with intervening valleys of varying breadth and depth. The bed of the St Lawrence is of a similar character. The channel runs along the strike of the upturned metamorphic beds, and consists of alternate ridges and hollows, as the dry land does. Where the ridges are elevated they form islands, rocks, and longitudinal reefs; while the valleys form the channels along which vessels proceed.

But from this description it will be understood that the deep water-channels, formed by the washing away of the softer parts or beds of rocks, must, such as these are, be irregular in width and direction. Where the river is broad, there may be several channels or parallel longitudinal valleys, any one or all of which may be interrupted by narrow and shallow reefs, which will render navigation intricate, difficult, and—when the tides run with great velocity—dangerous.

Such is really the case in the St Lawrence. About five miles below Quebec, the Isle d'Orleans divides the river into the north and south channels, and beyond this island, which has a length of twenty miles, it is divided into three irregular—the north, middle, and south—channels, by parallel ridges, the highest points of which form islands, and the lower, rocky or sandy reefs, visible only at low-water. Shoals also, at various points, stretch out from the south shore, which narrow and give still more intricacy to these channels. Hence, at a place called the Traverse or Narrows, about fifty-five miles below Quebec, though the river is there thirteen miles wide, the channel usually selected by pilots is only 1800 yards in width, and, to add to the difficulty, the ebb-tide runs through it at the rate of seven, and the flood of five or six miles an hour, and there is no anchorage.

From such circumstances as this arises the risk usually understood to attend the navigation of the St Lawrence, and which is one cause of the higher rates of insurance usually demanded for vessels which sail to or from this river.

The mouth of the river also has its dangers; and among the consequences of these, the most distressing are the shipwrecks which occasionally occur, after the winter season has set in, and when unhappy crews, having saved their lives from the sea, are thrown upon frozen shores or islands, far from human sympathy, or necessary supplies of food, fuel, or clothing.

To obviate the dangers arising from the intricacy of the navigation, lighthouses have already been erected at various points; but the number of these is still insufficient; and a people who have expended the large sums I have mentioned in improving the upper parts of the river cannot hesitate—now that their sacrifices are beginning to be appreciated, and are likely to meet with their reward—to organise and maintain a sufficiently extensive lighthouse department, to give confidence and security to the navigator.

It is not more on behalf of humanity, than as a matter of wise economy, that I would suggest the establishment of fixed depôts of provisions, and other stores, in charge of the necessary number of people, at different points on the islands or coasts about the mouth of the St Lawrence, where shipwrecks most frequently occur—that those appalling evils may be averted, which to us, who live in more genial climes, appear among the most fearful to which castaway mariners can be subjected. The expense of such winter depôts would be only small, while they would create in the breasts of seamen such a feeling of confidence as would often prevent the disasters, the consequences of which they are designed to relieve.

No legislative interference, of course, can ward off icebergs from the banks of Newfoundland, or make the seas more safe in the bay of the St Lawrence, in early spring, and when winter approaches; but greater skill and care in the masters of vessels may lessen the casualties arising from these sources. The proposed examination of master mariners, the advantages of which have already been frequently discussed in the Home Parliament, will go some way towards securing this greater care and skill.

There is still one branch of the internal navigation of Canada which is likely to tend not only to the extension of the traffic on the St Lawrence, but to the improvement also of the general commerce of the province.

I have already spoken of the river Richelieu as flowing from Lake Champlain, in a northerly direction, through the once fertile flat country in which Chambly, St Hilaire, and other villages stand, and out of which rises Belœil, and the other isolated mountains, which add so much to the picturesque character of the district. This river falls into the St Lawrence at Sorel, forty-five miles below Montreal. From this point upwards to Lake Champlain, the Richelieu has been made navigable by the lock or dam of St Ours, and the canal of Chambly, extending a distance of eleven and a half miles from the town of Chambly to St John, between which places considerable interruptions occur in the bed of the river. This canal cost £120,000; and though it has hitherto returned comparatively little revenue, the state of the trade on the Erie Canal is likely very soon to give it an important part of the traffic between the western country and New York.

I have already adverted to the excessive crowding of the Erie Canal, and the delays to which merchandise is occasionally in consequence subjected. But to descend to Montreal and Sorel is easy, and can be done without transhipment; and, in consequence of this and other advantages, it has been found that goods can by this route—down the St Lawrence, then up the Richelieu to Lake Champlain, and thence by canal to the river Hudson—be carried to New York as cheaply, and with more certainty as to time, than by the hitherto exclusive line of the Erie Canal. It may, therefore, be confidently predicted, that a portion of the internal traffic of the States will hereafter pass by the river Richelieu, and enrich and increase the value of land in the district through which it passes.

A shorter ship-canal has also been projected direct from Caughnawaga—opposite to Montreal, but above the rapids—direct to Lake Champlain. Should this be executed, there can be no doubt that much of the traffic between the western regions and the Atlantic borders would pass, without changing bottoms, in this direction—greatly adding, of course, to the income of the provincial canals, and to the commercial establishments and intercourse along the river.

On the whole, therefore, it appears certain that the river St Lawrence is destined ere long to become a most important medium of intercourse between the various sections of the New World, as well as between the Old World and the New, and to give to the province of Canada a far more extensive and commanding influence over the commercial operations of North America than any State east of Louisiana can ever aspire to.*

* I am happy to learn that, so far as the present year 1850 is concerned, the anticipations of the text have been fully confirmed. The revenue

I do not in this place introduce any remarks in relation to the railroad communications with the Atlantic which are now projected or in progress, as the observations I subsequently made in New Brunswick will naturally suggest some considerations in connection with this important means of colonial development.

from public works, during the first ten months of 1849, was £64,601, while, during the same ten months of 1850, it has been £76,672, an increase of nearly one-fifth. This, with the other abundant symptoms of prosperity experienced in the colony during the present year, will, I hope, hush the cry of discontent and disloyalty for some time to come.

CHAPTER XIV.

Leave Mitis for the Restigouche.—Nature of the road through the forest.—Clearings and accommodation by the way.—Mountainous character of the country.—Frazer's.—Elevated table-land.—Great Metapediac Lake.—Rich flat land around it.—Brechut's.—Little Metapediac Lake.—John Low's.—Rough road.—Dark night.—Burned forests.—Burned bridges.—Cause of the burnings.—Their effect on the landscape and on the soil.—Noble's.—Evans's Hollow.—Solitary life in the forest.—Hardwood ridges.—First green fields and clearings.—Home thoughts and associations.—Scottish settlers twelve hundred feet above the sea.—How such spots become known at home.—Beautiful scenery.—Fine land and farms.—Dixon's.—Abundant wheat a thousand feet above the sea.—Yankee phrenologists, soapmakers, and other adventurers; their luck in the provinces.—Campbelton.—Mr Ferguson of Atholl House.—Changes in a new country during a single lifetime.—River Restigouche.—Good land on either side.—Excursion up the river.—Flat lands.—Views on the river.—Scottish settlers from Arran.—General prosperity of these settlers.—Attachment to home recollections.—Goatfell.—Ilfracombe and the Causeyside. — Good land on the Upper Restigouche. — Alleged home ignorance of provincial geography. — Similar real ignorance in the colonies.—Indian settlement opposite Campbelton. —Progress of the Indians in farming.—Their winter employments.— Want of a school.—Sugar-loaf Mountain, and the view from it.— Geological reason for the quality of the land.—Good land on the Canadian side, in the county of Bonaventure.—Agricultural societies in this remote region.—Encouragement given to such societies by the Canadian Legislature.—Agricultural show on the Restigouche.— Prosperity of the lumber trade on this river.—Historical recollections of the river and bay.—Old French settlements.—Town of Dalhousie. —Increase in the growth of wheat in this district.—High price of Canadian flour.—Fossiliferous limestone at Dalhousie.—Settlers on the Eel River.—Their prosperity.—Great success of the potato culture.—Supposed superiority of the lumberer over the farmer.— Greater value of the resident farmer to the province.—Effects of a failing lumber-trade on the permanent welfare of the country.—Illustration of certain social and domestic differences between the United States and the colonies.

October 6.—Having yesterday arranged with a habitant, bearing the illustrious name of Dumas, to convey me across the peninsula of Gaspé for £5 currency, I rose from a short sleep at one o'clock this morning, and, after a drive of three miles inland from Mitis—for a considerable part of the way over a bog, upon a fearful corduroy road—reached the house of Dumas. Having transferred myself and luggage into the waggon I was to occupy for the next two days, we ascended a hill which separated us from all further communication, even by sight, with the river St Lawrence, and in about half-an-hour had entered the forest. Under the shadow of perpetual trees we continued, from this point, for a distance of eighty miles, emerging only to come within sight of the river Restigouche.

This road between the two rivers is a very rude and difficult one. It is barely blocked out of sufficient width to allow a waggon with one horse to pass. The trees are cut down and hauled off, boulder-stones and small inequalities removed, and bridges built where they are absolutely necessary. Only the horses of the country, which all their lives have been trained to it, could conduct even light waggons across the numerous steep hills over which the road passes. I had been told in New Brunswick that the road was impassable for carriages, and that my portmanteau would have to be carried, while I walked on foot myself a considerable part of the way; and I did think that my luggage, my conductor, and myself were a very heavy load for the little Canadian horse, till I afterwards saw other horses compelled to drag at least twice the load along the same road.

The traverse from Mitis to Campbelton is usually accomplished, in these light waggons, in three days. By starting at the early hour of one in the morning, I was enabled to perform the entire journey by the evening of the second day, and thus to avoid the disagreeables of two ensuing nights of wilderness accommodation.

The houses met with in the forest are very few, and their relative distances nearly as follows:—

To Frazer's, a log-hut attached to a small clearance in the wood, is from Mitis about seventeen miles.

To Brechut's—a house and clearing on the Metapediac Lake. In this house there is one inferior bed for strangers. When I passed, it was inhabited by three men, without any female. A change of horses can sometimes be got here, but it is not to be depended on, and could not have been got at the time of my visit. Nine miles.

To John Low's—a log-hut and small clearing on the Little Metapediac Lake, inhabited by John Low and his wife—eighteen miles.

This man was about to remove to a new clearance he was making five or six miles nearer Brechut's; while a son of the latter was to settle in Low's house, on the little lake, so that future travellers will probably find more settlers along this road than it was my fortune to meet with.

To Noble's—a larger house, better provided than Brechut's, and where a night may be spent, as my conductor said, "*sans grand misère*"—eight miles.

To Evans'—a solitary resident in a log-hut, in a wild small hollow of flattish land, surrounded by high, steep, pine-timbered mountains, almost precipices, on the banks of the *White River*, as it is called by the Canadians—twenty miles.

To the nearest clearing from this towards the Restigouche, on fine hardwood land, twelve hundred feet above the level of the sea—twelve miles.

To Dixon's—a clean and comfortable inn kept by a Scotchman, Mr Dixon, surveyor of the road, mail-contractor, and the owner of eight hundred acres of excellent land, in one of the most beautifully picturesque situations in all Canada—four miles.

To the Restigouche ferry—a rapid descent from Dixon's to the river—eight miles.

The whole distance being about 100 miles, during 80 of which the road runs through a continuous, almost untraversed forest. With the single exception of the postman in his one-horse car, who passes along once a-week, we did not meet a single individual the whole way through the forest.

The first five leagues, as far as Frazer's, were exceedingly mountainous and difficult. I have already described the interior country of Lower Canada, bordering the St Lawrence, as consisting of a series of ridges running parallel with the river, and separated by intervening more or less narrow valleys. Such is the country as far as Frazer's. It is a prolongation of the New England mountains towards the promontory of Gaspé—a succession of steep climbs and difficult descents, the latter being often not less painful and oppressive to the horse than the former.

Frazer's is situated at the commencement of a table-land, which extends for nearly three leagues, forming the best and most easy part of the whole Kempt Road. From this table-land, a long gentle descent brought us down to the level of the Metapediac Lake, on the edge of which stood the house and farm of Brechut.

Around this lake there is much flat land—generally, so far as I could see, either a stiffish light-coloured clay, or more or less soft, wet, and black swamp.

All, however, is covered with wood, with the exception of some of the drier parts at the one end, which Brechut has cleared and brought into cultivation. Blocks of limestone are scattered about the shores of the lake; and the same rock, resting upon a white sandstone, occurs *in situ* along its south-western border. This limestone contains fossils, and is considered, by Mr Murray of the Canadian Geological Survey, to be an extension of the limestone formation which, with intermixed, often greenish calcareous shales, forms stupendous sea-cliffs, 700 feet in height, at Gaspé Promontory, and, altogether, attains a thickness of 2000 feet. I had not an opportunity of ascertaining whether the age of this limestone had been exactly made out by Mr Murray, but the fossils he mentions indicate that this extensive calcareous formation of Gaspé is closely related to the Helderberg series of the New York geologists.

If such be the age of this limestone, a belt of very good land ought to extend in a north-east and south-westerly direction, from Gaspé Promontory, through this part of Lower Canada. From what I saw of the borders of the lake, it appeared to me certain that, wild as it now looks, and remotely as it is situated, the time will yet arrive when drainage and the use of lime will make fertile wheat-land of the flat country which fringes this extensive sheet of water. A natural outlet for its produce exists down the Metapediac River, to the Restigouche on the south; and should the road be improved towards the north, by following the course of the streams instead of crossing all the ridges, as the Kempt Road does, the access to the St Lawrence will be made at least equally easy. A grant of 50,000 selected acres here now, would be a fine fortune for a family some three generations hence.

About nine in the morning we arrived at Brechut's, and, after breakfasting, and resting our horse for an hour, we started again for John Low's, on the small or lower lake. This part of the road is not beset by so many high hills and steep descents as between Mitis and Frazer's; but it is in many places nearly overrun again with a natural growth of young trees, and the roads are deep, stony, and full of holes, so that our progress was slow, and *shaky* in a very unpleasant degree. The borders of this lower lake are not void of beauty; but it is a solitary abode for a small family, and neither man nor wife in this hut looked as if they were spending a happy life.

From Low's to Noble's, a distance of eight miles, though not hilly, was the most disagreeable of the whole traverse. After a mile of tolerable road, came two miles of the roughest and stoniest I ever travelled. Every yard had its own jolt and shake in store for us, which sixteen hours upon a hard seat had not prepared me, at least, to undergo with a great degree of equanimity. And when we were clear of this bed of stones, and were beginning to mend our pace a little, the sun set, and the brief twilight had already nearly passed away, when we came upon a gloomy, miserable-looking burnt tract of land, sloping rapidly towards the Metapediac River, blackened by thousands of charred stumps, and intersected by numerous deep narrow valleys, often mere gullies, with little streamlets descending through the dark peaty soil which covered their bottoms.

As we almost groped our way over this melancholy tract, we found ourselves suddenly upon one extremity of a bridge, the other end of which yawned like a black gulf before us, and proved to have been burned. The eyes of my Canadian companion were fortunately better than my own, and the horse was pulled up in time to prevent our being all precipitated into the brook below. Having dismounted, we found an awkward steep descent to the edge of the water, and by means of a temporary bridge of logs, and then a steep climb, we gained the road on the opposite side of the ravine. I was congratulating myself that this difficulty had been got over before the darkness had become more intense, when we arrived at the top of a much steeper bank above another burned bridge, and had again to grope our way on foot, down one precipitous bank and up another—a labour which, with the waggon behind it, our active and willing little horse, much to my surprise, though with considerable difficulty, accomplished. It was the steepest breasting of a bank by a horse with anything behind it I have ever seen, either before or since. It was a help to us, in these difficult circumstances, that the bridges along this road were all painted white, a circumstance known to my conductor, and which enabled him to discover, before we came to the burned places, that something was not as it should be in advance. I was not sorry, however, when, in about half-an-hour from this last bridge, through the midst of to me impenetrable darkness, the sound of a dog barking, and soon after the gleam of a blazing fire through an open door, announced that we had reached the end of this long day's journey.

The burning of the forests through all this country, along the borders of the Metapediac River, not only renders it black and melancholy to look upon, but actually injures the surface-soil in quality, renders it incapable of easy settlement, and prevents the explorer from judging, by the nature of the timber, what the prospects of a cultivator on such land would be. Through the carelessness of lumberers, Indians, and other rovers in the woods, these forests have been burned over and over again. This want of care there is as yet no law to punish. The extraordinary heat and drought of the present summer have particularly favoured the spread of fires, and increased, beyond their average amount, the number of public casualties, which the provincial authorities are required to repair. Hence it is, I suppose, that, although the bridges I have spoken of were burned in June or July, the Canadian Government has as yet left them unrepaired. It is in winter, also, that large timber for such works can be most economically cut, and hauled from the place of its growth to the spot where it is required.

When this road was first opened and made a mail-route, this station of Noble's and that of Brechut, on Lake Metapediac, were fixed upon by the Provincial Government, and the two present occupiers placed in them, with small pensions, to keep open house for strangers and to facilitate the weekly progress of the post. This pension, much to his dissatisfaction, as he was at pains to inform me, had been withdrawn from Noble, but he still lingers in his wilderness home, surrounded by a flourishing family, apparently ignorant of serious privation, and receiving and accommodating gladly any traveller who is willing to pay—a quality the want of which, he assured me, was always to be suspected when the guest had French Canadian blood in his veins. I found myself therefore very comfortable under his roof—as travelling goes in the wilderness. A huge log-fire in the kitchen, tea I had brought with me from Quebec, fish from the Metapediac—which flowed past the house—potatoes from Noble's garden, new bread baked on a hot plate for the occasion, and finally a clean, and not uncomfortable bed, wound up the adventures of the day.

Saturday, Oct. 6.—Rested and refreshed, I arose at five; and by six o'clock had breakfasted, and was again on my way—as from Noble's to the ferry on the Restigouche, I was assured, was a good twelve hours' journey.

From Noble's to the solitary hollow—for it cannot be

called a valley—in which Evans was located is twenty miles. Several very long hills, and many very steep ascents and descents of a shorter kind, were passed during these twenty miles. As in the latter half of yesterday's drive, the soft or pine and birchwood forests, generally unbroken, still accompanied me; though here and there patches or ridges, more or less broad, of hardwood or broad-leaved trees occurred, which will hereafter be selected as centres for settlement, and from which the natural expansion of a growing population will gradually spread the arts of rural life over less promising districts.

This morning there had been frost enough to produce ice half-an-inch thick, and to make me, in addition to a Canadian home-spun coat over my English great coat, feel thankful for a Buffalo coat also, which my landlord on the Mitis had kindly thrust upon me for the journey. The hoar-frost which covered the woods through which we passed lingered upon the trees, and made the air around us chilly till far on in the day—when the sun became high enough at last to peep into the narrow forest channel through which we were sailing, and restore the green colours again to the snowy leaves.

Evans—solitary, and still young—we found digging and pitting potatoes, of which, for a single man, he had a good supply. He had also some oats, which were already cut. He had been a lumberer, and, like his companions, given to drink when occasion presented. In their society, he found he could neither alter his habits nor accumulate money. "I therefore foreswore liquor," he said, "and settled in this lonely spot." For four years he had lived here, and though he had liquor in his house, as he showed me, yet for four years he had abstained. In the autumn, he made a little money by going to the settlements to aid at the harvest. But the solitude had at last become too intense for him, and he was about to remove twelve miles nearer the Restigouche, to a ridge of good land where there were already many settlers, and where the sounds and sights and sympathies of civilised life would be within his reach. Of strong-minded men, like this Evans, capable under emergencies of great self-control and self-denial, these wilderness countries exhibit to the traveller many striking examples.

In Evans' hut, I made tea for the whole trio; and then, having fed our horse with a bottle of his new oats, we again proceeded. Twelve miles beyond, the country changed. It became broken with distinct hills and ridges. The forest trees altered also. Instead of pine and white birch, maple, red birch, and hornbeam, mingled with rarer pines, hung out their broad leaves—already assuming their autumn tints—to the declining sun. We had come upon what was called a hardwood ridge of strong fertile land; and I cannot express the welcome feeling with which, while still two thousand feet above the sea, I looked far forward along the narrow road, chiselled as it were out of the old forest, upon a green field in the distance—the first symptom of our approach to a region where art and human intelligence were successfully striving with the long mastery which unbridled nature had been exercising over the submissive soil.

We had some time before passed the culminating point of this extensive traverse, and were now descending a long slope looking towards the south. This descent brought us, in what I now thought a short period of time, to the final limits of the primeval forest. I almost felt unwilling to leave behind me its more cheerful broad-leaved beauty, as we emerged among cleared fields, where the still golden stubble indicated that wheat, or oats, or rye, had recently been reaped; and where the large green leaves of the turnip, and the blooming tops of the potato, still covered the ground.

After emerging from eighty miles of scarcely broken forest, it was to me a very interesting sight, upon the first green spot of fenced land which caught my eye, to discover a field of dark green broad-leaved turnips. Home thoughts spontaneously awoke with many home associations. Home pictures sprang up before me, and home culture and home improvement were called to mind by the sudden sight of these healthy and luxuriant crops of what, without a pun, may be truly called the *root* of agricultural improvement. Home breeds of stock also, though mixed—the sound of Scottish voices—a Scottish ploughman behind a Scottish plough—and females, not disdaining to assist in raising the potato which a kind Providence had this year preserved for them—all spoke of my own land, and of those rural districts where educated intelligence and contented industry unite in spreading the happy results of better husbandry. Even the noisy and troublesome cur, rushing out with his boisterous salute, was welcome to us, as we emerged from the soundless woods, where three or four rare birds were the only wild things we had seen in all our journey. And yet the autumn-tinted maple-groves, bordering the green fields, gave a peculiarity to the landscape which commingled foreign with home feelings, and kept still present before me the realities of my actual position. Along with all these pleasant imaginings too, it was a great relief to think that the more uncertain and unpleasant part of my present journey was over, and that for some time to come I might hope to to see the accompaniments of civilised life, and the evidences of human skill and industry around me.

These first settlements we came to are about eight miles north, in a straight line, from the banks of the Restigouche River, and 1250 feet above the level of the sea. That the crops and culture and farming I saw here should be possible at so high a level, shows, not only that the land is naturally good, but that this northern climate must be far more propitious to vegetation than is generally believed.

One thing the traveller through a region like this is surprised at, when he stumbles on a settled and cultivated tract of land, such as I was now passing through. He wonders how the people came to find it out. Who induced all these men and women to leave remote corners of Scotland, and settle in this remoter corner of south-eastern Canada? The whole line of country is a *terra incognita* at Quebec and at Fredericton. At the seats of Government of both provinces, where they complain of how little we know of their geography at home, the spot I speak of was absolutely unknown; and yet humble Scotchmen and their families had made choice of it, and already fixed upon it their future homes. There is an under-current of knowledge flowing among the masses, chiefly through the literary communications of far distant blood relations, of which public literature knows nothing, and even Governments are unaware.

To Dixon's, which is four miles from the first clearing, we passed through a succession of new farms—for all this tract is newly settled—for the most part upon good land, and descended about 200 feet, being now on the slope towards the Restigouche. On approaching his house and farm, which are still 1000 feet above the level of the sea, the view of the mountainous regions towards the south and west was truly beautiful. Trap rocks or mountains, in this direction, rise up among the stratified deposits, and give a new and magnificent character to the landscape, as seen from this comparatively lofty elevation. An amphitheatre of peaks stretches almost as far as the eye can reach; while, in the foreground, single elevations cheered the eye with the warm tints of the hardwood growth that covered them. I regretted that I could not spare even a single hour at Dixon's, to enjoy more fully the passing mountain scenery which glided so rapidly before my eye, as the carriage hastened on. On leaving St John, six weeks before, for my tour through western New York and Canada, I had made an appointment to meet my two New Brunswick travelling companions this very evening, on the shores of the Restigouche; and I had still eight miles and a ferry to cross, before I should arrive at Campbelton, the end of my journey.

From Dixon, I obtained most favourable accounts of the quality and productiveness of the land at this high elevation of 1000 feet. He is the possessor of 800 acres, and farms what is cleared of these. Besides turnips, potatoes, and green crops generally, for which his land is admirably adapted, he grows wheat yearly without fail. His wheat ripens well. The wheat-midge has never visited him as yet, though he is occasionally troubled with rust; and he has reaped upwards of 50 bushels an acre. Indian corn, he said, would ripen with him in such a season as this. It is to be hoped that, in clearing this fine district of country, the necessity of shelter will not be forgotten; and that the certainty of the harvests, now that the fields are surrounded by the native forests, may not be sacrificed by laying them too open to the cold winds from the Bay de Chaleur.

At Dixon's, I met with a Yankee phrenologist, and a Yankee maker of daguerreotypes, who, after a successful campaign in New Brunswick, were so far on their way to Lower Canada, to experiment on the heads and faces of the habitants along the shore of the St Lawrence. A certain number of these peripatetic philosophers find the British provinces a profitable country to explore. I dare not venture to put down the number of dollars I afterwards learned that these men had carried with them from Campbelton, the little town to which I was going. Such men as these not unfrequently hunt in couples;

and the French Canadians are capital game for them. To reach this race of men, they rarely come by the remote route which these two were taking. By Lake Champlain the road is shorter from New England, and the heart of the country more easily accessible. While I was on my short visit to St Hilaire from Montreal, two of these experimenters approached the St Lawrence through the eastern counties, by way of St Hyacinth and the railroad. One of them, for two dollars, taught how to convert a barrel of flour into a barrel of soap, in ten minutes! and the other sold, at a dollar a gallon, a black varnish, which was the best in the world. We are here in England as gullible as any nation on earth; but the gull*ers* generally arise from among ourselves. It may not be, therefore, that a better education would protect the Lower Canadians much from this class of impostors. They are far out of the world, and simple, because they are so; but education is a good thing in many other ways, and it may be fairly tried as a giver of greater practical wisdom too.

Dixon himself drove me down the last eight miles to the ferry, which I reached just as the twilight ended. Half-an-hour more took me across to the town of Campbelton, where I arrived about 7 P. M., having finished this journey in six days, as I hoped to do when I left Quebec on Monday morning. Hotels are not abundant, but I obtained tolerable quarters for the night; and on the following day, through the hospitality of my New Brunswick friends, was provided with excellent accommodation.

Monday, Oct. 8.—The only church at Campbelton is a Presbyterian, and the clergymen being from home yesterday, there was no service. I therefore walked up the river two or three miles to Atholl House, where I was induced to remain for a couple of days under the hospitable roof of Mr Ferguson. To all who have ever visited the Restigouche, this gentleman, by character at least, is known. He was one of the first British settlers in this part of the colony, and one of its most energetic improvers and explorers. Mrs Ferguson was the first child of British parents born upon the Restigouche. Where there are now so many old farms, and comfortably settled inhabitants, it was very interesting to find the first explorers still alive, and witnesses of the pleasing results of their early exertions.

The Restigouche is here the boundary between Canada and New Brunswick. Flowing from the west, it rises partly in the Canadian and partly in the New Brunswick highlands. It receives various tributaries, of which the Upsalquitch is the most important from the New Brunswick, and, six miles lower down, the Metapediac from the Canadian side. Soon after the junction of the latter stream, the Restigouche widens out into a spacious harbour about two miles in breadth, and four-and-twenty miles in length. The mouth of the harbour or river is at Dalhousie, sixteen miles below Campbelton, and it is deep and navigable for the largest ships almost as far as the tide ascends, which is six miles above Campbelton. It is bordered on either hand by a belt eight or ten miles broad, of excellent hardwood upland, resembling in quality that which I had passed through on the Canadian side after I had emerged from the forest. Along the river are margins of flat intervale, sometimes narrowing to a mere fringe, at others expanding into fertile alluvial tracts containing hundreds of acres. Of the upland, the greater portion is still under virgin forests; but the flat lands, for a long way up the river, are granted to actual or intending settlers, aud are more or less under cultivation. Below Dalhousie the harbour widens further into what is called Restigouche Bay, and, finally, into what the early French discoverers named the Baie de Chaleur.

We devoted this forenoon to an excursion up the river six or seven miles, to what are called the *Flat Lands*, *par excellence*. These consist of about five hundred acres of alluvial flats and terraces, which skirt the river chiefly on its left bank, forming fine arable farms. The day was fine, the air clear, and from some of the hills which skirt the road along the river, the view was very beautiful. At the head of the tide-water, where the harbour narrows into the river, its bosom is studded with upwards of twenty small islands. These little sunny wooded islands immediately beneath our feet; the interminable river stretching upwards, now seen, now lost amid the hills and forests; the far view to the right, carrying the eye beyond the harbour and river-bay till it lost itself towards Miscou Island, where the waters of the Bay de Chaleur intermingle with those of the Gulf of St Lawrence; and more near, the Sugar-loaf Mountain lifting its solitary bulk on the right bank between us and Campbelton; while lower down, on the Canadian side, the loftier Tragadegash, wooded to the very summit, towered over the entire channel: all this, in the clear sunshine of this climate, formed a delightful picture. And the beauty of this picture was heightened by the frame-work of high lands on each side, between which it was all enclosed. Enlivened by the autumnal tints which characterise the hardwood forests of North America, the mountain-ranges on either hand seemed to rejoice in the bright warm rays from the sunny sky, while they spoke to the instructed observer of agricultural capabilities in the yet untouched wilderness, which we Europeans are little accustomed to look for in so remote a region as this.

The settlers on the front Concession along the river, above and below Campbelton, are chiefly from the Scotch island of Arran; and they are all thriving—not laying up money, but independent—and bringing up their families, which are usually large, in comfort and plenty. The progress of settlement in this promising district may be judged of from the fact that, six years ago, the front Concession or row of farms was scarcely occupied, whereas now there are settlers as far back as the third Concession. And yet there has been no rush of emigration to this point, the arrivals of new settlers being small and gradual, and rather of a *dropping* character. The back-settlers are chiefly Irish. Those of this nation who do come here appear generally to thrive, as there is not a single pauper in the whole large county of Restigouche, with the exception of a straggling French Canadian now and then, from the opposite Gaspé country, on the Bay de Chaleur, or an unsteady and idle Irish immigrant.

I went into several of the houses, generally of small pretensions, of the settlers on the banks of the Restigouche at this upper part. They were Scotch, English, and Irish. All expressed themselves as being happy and contented, and their children looked healthy. Very old people abounded, which also spoke for the health of the climate; and it was said to be rare for children to to die young.

Few things are more interesting in a strange and distant land—carrying you sooner into the hearts of the people, and giving you with them the position and familiarity of an old friend—than to be able to talk to them about their old haunts *at home*. In one cottage the mistress was now a widow: she was from Devonshire, and had been many years before a servant of the incumbent of Linton. I spoke to her of Bideford and the Valley of Rocks, and Ilfracombe. It was holding up to her a picture of old and happy days. "Oh sir!" she said, as I left her, "I do so like to hear about Ilfracombe and Combe Martin, and all them places."

But a broad Lowland Scotch tongue, and a knowledge

of Scottish localities, will make a man at home in a greater number of houses in New Brunswick than almost any other qualification which a Briton can possess; and I think I spoke more broad Scotch during my three months' tour in New Brunswick than I had done during twenty years of my life before.

On my previous tour upon the St John River, as we were driving through a new settlement, a farmer and his staff, who were cutting oats, stopped to look at us. I was told he had come from Paisley, so we pulled up to talk to him. "Would you raither be staunin there, or at the corner o' the Causeyside?" I said to him. This unexpected allusion to his native place went straight to his heart. He stood for some time without reply, and then said,—"Ah, sir, the Causeyside's a bonny place." Those who know the kind of beauty possessed by the Causeyside of Paisley will understand how much heart and home affection was expressed by this word "bonny."

Among the Arran settlers on the Restigouche, the love of country which bound them to their island-home has been transferred to the similar land of "mountain and flood" in which they are located. After other lively talk with a middle-aged thriving farmer, and comments on the country, and comparisons with home,—"An' is na that hill like Goatfell?" pointing to the lofty Tragadegash on the opposite Canadian shore. He could scarcely express his assent; and after our conversation was ended, and I and my friend had entered the carriage, he came warmly forward with his outstretched hand,—"I maun hae anither shake o' yer han', sir; ye're a real Scotchman."

Above the flat lands there are settlers scattered along the river some ten miles farther up, as far as the mouth of the Upsalquitch, which I have already mentioned as an important tributary of the Restigouche. The flat lands at the mouth of this tributary are chiefly on the Canadian side; and there Mr Ritchie, the most extensive lumber-merchant on the river, has a farm on which he has this year raised five thousand bushels of oats, and where green crops are said to thrive well. This gentleman informed me that, about fifty miles higher up, the river passes for ten or twelve miles through an undulating tract of land of considerable breadth of the richest quality, and, as he thought, one of the most promising in the province. His lumbering expeditions gave him opportunities of seeing and judging of the country not possessed by other parties; and I insert these and other particulars regarding this river because it struck me, from its natural beauty and fertility, and from the peculiarly healthy tone of character displayed by its present rural population, to be more worthy of the attention of those desirous of changing their homes than either we or the New Brunswickers generally are in the habit of supposing.

As I have above remarked, the people of the colonies sometimes express great indignation that we at home know so little about their capabilities, or even their geography. But my own experience in New Brunswick—which province I explored the most thoroughly—is, that the provincials themselves know their own country little better than we do at home. I may even doubt if they know some parts of it so well. But supposing our ignorance as great as some incline to represent it, this answer to the complaint might be made. The whole of the vast territory of New Brunswick contains little more than half the population of the cities of Manchester or Glasgow. A single street in either city contains more people than the majority of their towns and cities. As well, therefore, might the householder in an obscure lane in any of our great cities complain that his countrymen did him wrong in not making themselves familiar with the local geography and whereabouts of his humble residence, as the colonial villages grumble that they have no place in the geographical recollections of the people and Legislature at home.

At the Governor's table one day at Fredericton, I sat next to a lady, said to be a great heiress, the daughter of an Englishman born, who was complaining loudly of the little that was known of their country at home. "Allow me," I said, "to try your geography? Is Ireland to the east or west of Great Britain?" She could not answer me; and it was unnecessary for me further to defend, on that day at least, our home knowledge in geography against the attacks of the provincials.

In the afternoon I crossed the harbour from Atholl House to Mission Point, to visit the priest, Mr Olscamps, and the Indian settlement he has charge of upon the adjoining land. This is a settlement of Micmacs, which the priest has very much the merit of keeping in peace and sobriety; and to his exertions it is mainly owing that the Indians have been induced to settle upon the land assigned to them by the Government, and to attend to the operations of farming.

Ninety-four families, comprising in all 410 individuals, are here located upon 800 acres of flat land, lying between the mountains and the river. Five years ago, when Mr Olscamps came among them, they were nearly all drunkards; but he has succeeded in greatly improving their habits and condition since his arrival. He speaks their language, and had previously been four years among the Indians of the Hudson's Bay territory, and had learned the art of influencing them. There are still many idle ones among his flock, and many who will not settle to regular labour, preferring hunting and fishing to steady work upon the land; but the greater number cultivate their small farms, and raise potatoes, oats, beans, and Indian corn, sufficient for their own consumption.

I went into the houses of some of these Indian farmers; and though they had still the habits and peculiarities of their race, I found them industriously engaged with their potatoes and Indian corn. The largest cleared farm held by one individual was 30 acres; another had 25 cleared; a third 20; many 15, 10, and so on, down to 2 or 3 acres. One of the chiefs I visited had 20 sheep, and others had smaller numbers. The greatest difficulty of the *curé* lay in his inability to prevent the people of Campbelton, on the opposite side of the river, from selling *fire-water* to his thoughtless flock.

In winter they work in the woods, cutting firewood or lumbering, as many of the young men are excellent axemen, and are in request as woodcutters. The older men hunt martens, and sometimes make considerable sums of money in this employment. Many prefer wintering in the woods with their families to living in their houses. At home, indeed, we have no idea of the superior warmth and comfort found amid the shelter of the woods during the winters of these northern countries.

Maple-trees abound on the hardwood land of this region, so that in spring all go to the sugar-making. Some families make from six hundred to a thousand lb., which they sell at 5d. to 6d. per lb. They require all this money to buy necessaries, as they do not as yet raise food enough for their own consumption; and in spring, flour sells at Campbelton and Dalhousie as high as eight or nine dollars a barrel.

There were four ploughs in the settlement, all of good construction. On one of them I saw irons with the name of "Wilkie, Uddingstone, near Glasgow," upon them. I sympathised with the wishes expressed by the priest to be able to maintaiu a permanent school among them. Difficulties stand in the way of a Government grant, inasmuch as, by the Canadian regulations, aid is given only to schools which have a certain minimum number of pupils. It is in forming the school, and breaking in the children to attend at all, that the main difficulty lies. When formed, it might possibly be made self-supporting. So far as I have learned, all the provincial Governments have been anxious to minister, as far as their means and knowledge go, to the comfort and improvement of these vanishing races of people; though individual feeling and party bias in the provinces, as at home, occasionally obstruct for a time the adoption of the most judicious and beneficent proposals.

Tuesday, *Oct.* 9.—After breakfast this morning we climbed the sugar-loaf mountain—a warm, difficult, and steep ascent over rocks, stones, and *windfalls*. It is upwards of eight hundred feet in height, though not high enough, as I had expected, to give us a view over the uplands towards the south, though these are not so high as those on the Canadian side. The prospect from the summit was extensive and beautiful. Up the river, the eye penetrated as far as the gorge through which the Metapediac enters the Restigouche from the north, and down—beyond Campbelton and Dalhousie on the one side, and the Tragadegash Mountain on the other—far along the Bay de Chaleur, almost to its mouth. North and south, hardwood ridges, with intervening valleys, extended as far as our elevation permitted us to see. It was clear that, for many miles on either side of the river, the land was naturally favourable to agricultural settlements, and partook in a great degree of the good qualities of the soils I had passed over on my way from the Canadian forest to the ferry on the Restigouche.

It is interesting to observe how geological structure is connected with and explains all this. Over the limestone formation, which is observed about the upper Metapediac Lake, lies a thick deposit of sandstone, which forms the surface of the pine-clad country I passed through for a considerable distance on my way towards the south. To this succeed a series of beds of a more mixed, shaly, calcareous, and sandy nature, which form the improved hardwood lands that border the shores of the Restigouche River, harbour, and bay, and of the broader Bay de Chaleur. They comprehend the representatives of the Devonian and mountain limestone systems, though, in a new country like this, so little explored, and, from its covering of forest, so difficult to explore, the limits or details of these two formations, or of their subdivisions, have not as yet been made out. But it is to the presence of these formations that the good land of this region is to be ascribed, aud by their extent, in a great measure, that it is limited. Climate, therefore, unless it be extreme, is by no means the most influential element in determining the agricultural capabilities of a country. Its geological character has still more to do with its economical prospects, and is deserving of a study not less careful and minute, both by natives and foreigners, than is usually given to climatic conditions.

The part of Lower Canada to which I have so often alluded as forming the northern shores of the Restigouche and of the Bay de Chaleur, constitutes what is called the county of Bonaventure. In this county, especially along the shore towards the east, there is much good land, many villages or towns, intelligent settlers, and extensive settlements. From all I saw and learned, I believe it to be one of the most favourable parts of Lower Canada for the homes of British settlers. It is far from the seat of Government, and on that account more likely to suffer neglect; but it is somewhat nearer to us, and more accessible from our British ports, chiefly through vessels which bring the yearly supplies of timber to the British islands.

Agricultural improvement is not unthought of even here. I crossed over this afternoon to the north side of the harbour, to be present at an agricultural show for the western part of the county. There were altogether 130 articles exhibited—cattle, some Durhams, but chiefly Ayrshires; sheep, and horses, which were very creditable to so new a country. The potatoes, turnips, and cabbages, &c., also were all excellent. It is complained, as in some parts of the States, that the market for mutton does not increase so fast as the production of sheep. The societies, therefore, are beginning to agitate the propriety of encouraging the introduction and rearing of merinos, for the sake of the wool, or, in the meantime, as more immediately attainable, of a cross between the Leicester and the South Down.

If not always a sure indication of progress, this existence of agricultural societies in all these remote places is a sign that, in the minds of a certain number of persons, there exists a desire to progress. There are two of these societies in this remote Canadian county of Bonaventure, and others in the county of Gaspé. The Canadian Legislature add £3 to every £1 which is subscribed by the members of a society, provided that the whole sum given to the county from the provincial purse do not exceed £160 a-year. Thus, £25 raised on the spot secure £100 to give away in prizes. Each great district, also—of which there are four in Lower Canada—receives in succession the large grant of £500, which is given in prizes at the great quadrennial shows held in succession in these several districts. The district of Montreal has the advantage of this grant during the present year.

The soil at the foot of the hills in this neighbourhood is much infested with the *Galeopsis tetrahit*—the day-nettle, as it is called here. And the curious thing about it is, that the seeds of this weed are often turned up in hoards or nests, of handfuls at a time, especially upon new land. Some animal of course has collected them, and stored them away for winter food.

Crossing again to the New Brunswick side, I started for Dalhousie, sixteen miles down the river, and at the mouth of the harbour. Both Campbelton and Dalhousie are new towns. Twenty years ago there were only three houses in Dalhousie, and only one old house where Campbelton now stands. Both are already considerable places, and contain together upwards of 2000 of a fixed, besides the less stationary lumbering population.

The lumber-trade, which in so many other parts of New Brunswick has failed, and given rise to much discontent, was described to me as being on the Restigouche as prosperous as ever. But the mode of conducting it has been changed. Instead of making advances, as formerly, to persons who led out parties into the woods, and delivered the timber in spring to the merchant at a price, the merchant now engages his own gangs of cutters, places his foremen over them, provides their supplies, and the logs when they arrive are his own. Measures are taken also to diminish or do away altogether with the enormous commissions and agents' charges, which, on both sides of the Atlantic, have hitherto stood so much in the way of a steady and fairly profitable transaction of business on the part of the manufacturer, or of the merchant exporter.

The Bay de Chaleur, and Restigouche harbour and bay, are somewhat famous in the history of this colony. The French first colonised this river, and established forts and settlements upon it. British fleets and troops have fought in it, and the remains of French defences are still to be seen on many of the points most suitable for defence. Mr Ferguson preserves around Atholl House several large guns, which had been buried and left behind on the site of an old battery upon the Canadian shore, a few miles above his residence.

Oct. 11.—Immediately behind Dalhousie rises Challefours Hill, from which an extensive view is obtained, not only of the river-scenery already described, but of the newly-opened country up the Eel River, a stream which flows into the bay a few miles below Dalhousie. Up this river there are many new settlers, all Scotch at the settlement of Dundee, and a mixture of Scotch and Irish at that of Colebrook. These have all been located within these five or six years, and are all prospering. Every kind of grain ripens. Even Indian corn—the short eight-rowed yellow variety—is a sure crop. Only nine years ago, when the local agricultural society was established, it was believed that wheat could not be grown; now most farmers grow not only enough for their own consumption, but have some for sale. The variety sown is a hardy red chaffed spring wheat, with a long red beard, known as the Red Russian. This variety has hitherto escaped the midge, and is less subject to be shaken when over-ripe.

Still the lumberers, and the Indians, and the townspeople, and the new settlers require imported flour; and it shows how imperfect the means of transport still are about the mouth of the St Lawrence, or how scarce capital or mercantile competition, that when flour sells at Quebec at 20s. a barrel, it brings here 35s., and that on the closing of the river it rises here at once to 50s. a barrel. At this time of my visit, though still early in October, it is 40s. a barrel for cash at Dalhousie. One of the benefits of the harder times on the North American rivers will be, to wean the lumberers from their attachment to the finest and fairest flour. They have grown up under the idea that the darker flour produced from their home-grown red wheats was inferior in nutritive quality, and not good enough for their subsistence. The sooner all parties are disabused of this erroneous impression the better—the less of imported flour will be required in Nova Scotia, New Brunswick, and Lower Canada—the more self-sustaining they will become—and the more prosperous the agricultural interest of the colonies.

Oats imported from Prince Edward's Island are here selling at 1s. 8d. currency per bushel, the freight being usually 4d. a bushel from that island, and 3d. from Quebec. This grain is extensively imported from Prince Edward's Island to New Brunswick. Labour is said to be cheaper there—£15 to £18, with food, being the wages of men, who, on the Restigouche, would ask £25 to £35 currency. But labour is always higher where lumbering is carried on to any extent.

The old red sandstones are seen in a nearly horizontal position on many parts of this coast. Along the shore, about a mile below Dalhousie, an interesting cliff section is exhibited of highly inclined rocks, consisting of limestones and calcareous shales, full of fossils, intermingled with harder, somewhat metamorphic beds—altered possibly by the neighbourhood of trap-rocks, which also abound along the south shore of the bay. Among the fossils were abundant large madrepores, cyathophylla, and productæ, with tubipores, branched corals, delthyris, &c.; but whether the beds were upper silurian or mountain limestone, as they are coloured by Lyell—I suppose from Logan's survey—I had not leisure to collect fossils enough satisfactorily to determine. Any scientific geologist who may hereafter visit the Bay de Chaleur will find this an interesting point to examine.

This morning being fine, Mr Campbell of Dalhousie was kind enough to drive me up the Eel River as far as the newest settlements lately planted there. We penetrated to the very farthest log-huts at the extremity of the last new road driven into the forest. The settlers are not altogether without society; and when of the same country and religion, as they suffer alike, so they sympathise deeply with each other, and are ever ready to give a hearty and willing assistance to a distressed neighbour. Though not so close together as in the French settlements, where the long narrow farm system prevails, yet 80 or 100, or even 200 acre lots, do not separate people very widely, and the habit of erecting two log huts opposite each other on farms separated only by the road, gives one near neighbour at least in a district which is rapidly filling up. To the enjoyment of a fair measure of happiness, only a few intimate friends are necessary; and we who live among a thickly crowded population know little how strongly, in a wilderness country, the hearts of neighbours become knit together where those external influences of more civilised life are excluded by which differences and discontent are most frequently awakened.

The idea I have formerly alluded to as being entertained in New Brunswick, that lumbering—spending the winter in the woods cutting down trees with the axe, and the spring in guiding the logs and rafts down the swollen streams—is the more dignified occupation, and that the lumberer degrades himself when he becomes a farmer—such an idea calls up a smile on our faces at home. Such, however, was everywhere, till recently, the notion entertained by the active enterprising young men on the North American rivers. But bad times, and perhaps other influences, are correcting these notions, while, at the same time, they are improving their morals. They begin now to consume tea instead of ardent spirits. They thus save money; and seeing that the farmers—though they cannot show so much ready money as they are themselves sometimes master of—are, on the whole, more independent, and less liable to great vicissitudes than themselves, they now look to farming as a pursuit to settle down in, are buying land, spending their spring leisure in clearing it, and, when they have prepared it for their families, in placing them permanently upon it. Of this class of settlers are many of those whose farms I saw on my excursion up the Eel River to-day, and it is in this way that what has been called the failure of the lumber-trade—which, at the shipping ports, has made the merchants exclaim loudly against the change of the timber-duties at home, and has even put the cry of Annexation into the mouths of many—is really leading to the most permanently beneficial results for the colonies themselves. The steady settled farmer is worth, to the future welfare and prosperity of the colony, a dozen unsettled lumberers, who this season may cut timber on the provincial rivers for the merchants of St John or Quebec; the next may be off to the Aroostook or the Penobscot, in the service of the merchants of Bangor or Portland; and the third may be found in Georgia, toiling among the pine-barrens for the lumber-merchants of Boston; and who, if they remained in the colony, would continue an unsteady and unthrifty race. It is not denied that the mercantile interest has hitherto exercised, in some of the North American colonies, more than a due share in the management of affairs, and that past legislation has been biassed considerably by this dominant influence. It cannot be doubted that the same influence has also operated, through the press and the hustings, in creating, or endeavouring to create, an impression in the provincial mind as to the feelings of people at home, and as to the purpose and ultimate tendency of home legislation, which will not, I think, be justified by further knowledge and experience.

To show the natural capability of the new land these

men settle upon, and how, at this time of potato-failure, they have husbanded their seed-potatoes, I may mention the returns obtained this year by several of these settlers on the forks of the Eel River, as I took them down from their own mouths. One of them planted 4 barrels, cut into small sets, and dug up 142 barrels; another planted 3 barrels and housed 60, besides eating of the crop all summer; a third planted 3½, and took up 60, besides consuming, since the 1st of August, what were required by a family of six children three times a-day. These large crops are given by land on which the wood is cut in the fall, the trees burned in spring, the ashes spread, and the seed put in. The comparative greatness of the returns may be judged of by the fact that, in the highly farmed and highly manured land around Edinburgh, 4 bolls, or 16 cwt., are usually planted; and 6 to 8 tons (8 to 10 fold) are reckoned large crops to raise on the West coast. In Ireland 10 to 12 tons are frequent, but even this return is small compared to that of these necessity-compelled, thrifty New Brunswickers.

I went into the furthest log-hut upon the last clearing. It was warm and comfortable; and a good stove in it not only kept the inmates warm, but gave them the means of cooking. These cooking stoves are found very convenient in North America. Numerous varieties of them are exhibited at the larger agricultural shows, and some of them do their work with a great economy of fuel—an article which lavish expenditure in past years is beginning to make scarce and dear in many of the more densely-settled districts.

I found the wife and five clean healthy children in the hut. She was very content, and would not go home if she had an opportunity. It was foolishness which brought them away; and they are not so well off yet as when at home, but hoped in a short time to be better off, as I have no doubt industry would soon make them.

Potatoes, oats, and turnips were the crops growing on their first year's clearing—all luxuriant and healthy.

An anecdote told me by a friend at Dalhousie illustrates very graphically one of the most important of the social and domestic differences by which our own homes, and those of most of our colonies, are distinguished from those of the United States. "A settler of many years at Dalhousie, a shoemaker by trade, had saved £500 in money, and had five or six boys growing up, when he took it in his head to go off to Wisconsin. Six months after his departure, a small vessel from Quebec entered the harbour of Dalhousie, and, when evening came on, a depressed-looking man in shabby clothing landed from the vessel, and walked up to my house. When he came in, I was surprised to recognise my old neighbour the shoemaker. 'You are surprised,' he said; 'but though I was a fool to go away, I have had courage enough to come back. When I had got to Wisconsin, my boys — who had been good boys here — began to neglect their work, and disregard me. I durst not correct them, sir, or I should have been mobbed. They soon learned this, and my authority was gone. My heart was sore, my money was melting away, my children were a sorrow instead of a comfort to me, and talked of starting for themselves. I sold off and came down to Canada. "Now, my boys," says I, "I have got you under the British flag again, and we'll have no more rebellion." So I kept my boys in hand, but we didn't get on as we used to do; and, at last, I determined to come back to Dalhousie. What's the world to me, sir, if my boys are to be a vexation to me? But I haven't a penny of money; and our clothing is so scanty that I am ashamed to bring them all ashore in daylight.' *

* How different this picture of the domestic relations, in these new States, from the representations which have come down to us regarding the ancient republics of Greece and Rome! How different, for example,

"So I gave him," added my informant, "the use of a house of mine that happened to be empty; his wife and boys were brought ashore the same night, and they are again an industrious, if not so united a family as before."

from that fine old picture presented us by Cicero (*De Senectute*) of Appius Claudius, who, after being five years censor, having brought water into the city of Rome, and having built the famous Appian Way, had at last become blind, and retired into the bosom of his family. He thus makes Cato speak of the old man—"Quatuor robustos filios, quinque filias, tantam domum, tantas clientelas, Appius regebat et senex et cœcus. Intentum animum tamquam arcum habebat, nec languescens succumbebat senectuti. Tenebat non modo auctoritatem sed etiam imperium in suos, metuebant servi, verebantur liberi, casum omnes habebant; *vigebat in illo domo patrius mos et disciplina.*" Love, fear, and reverence were entertained towards the ancient father in that *old* republic—what better things can have taken their place in the *new?*

END OF VOL. I.

PRINTED BY WILLIAM BLACKWOOD AND SONS, EDINBURGH.

www.ingramcontent.com/pod-product-compliance
Lightning Source LLC
LaVergne TN
LVHW021144110826
845150LV00005B/1124
* 9 7 8 1 4 2 5 5 4 7 1 9 6 *